T0178611

Springer Biographies

More information about this series at http://www.springer.com/series/13617

Salvatore Esposito

Ettore Majorana

Unveiled Genius and Endless Mysteries

Translated by Laura Gentile de Fraia

 Springer

Salvatore Esposito
I.N.F.N.
Naples' Unit
Naples
Italy

ISSN 2365-0613 ISSN 2365-0621 (electronic)
Springer Biographies
ISBN 978-3-319-85369-7 ISBN 978-3-319-54319-2 (eBook)
DOI 10.1007/978-3-319-54319-2

Translation from the Italian language edition: *La cattedra vacante* © Liguori Editore 2009. All Rights Reserved.

Printed on acid-free paper

This Springer imprint is published by Springer Nature
The registered company is Springer International Publishing AG
The registered company address is: Gewerbestrasse 11, 6330 Cham, Switzerland

Acknowledgements

My interest in the scientific work of the protagonist of this book, Ettore Majorana, was first excited by Erasmo Recami some years ago. I take the opportunity here to thank most warmly for his enduring interest, encouragement, and kind willingness to make available the entire body of material he has collected, as well as his own personal experience. However, the English edition of this work would never have seen the light had it not been for Fabio Iocco's enthusiastic involvement: besides encouraging me to publish, he also handled the first contact with the publisher. The translation by Laura Gentile de Fraia then brought the project into existence.

Some interesting conversations with and suggestions by the Majorana family, particularly Ettore Majorana, Jr. (in Rome) and Fabio Majorana (in Catania), have been equally useful for the purposes of this reconstruction.

A significant fraction of the inquiries into Majorana's disappearance were carried out together with Roberto De Risi. I would thus thank him here for his invaluable collaboration, and also that of Gennaro Miele, on several occasions.

Many colleagues and/or friends have contributed, directly or indirectly, to the acquisition or interpretation of the scientific and historical material used in this work. Among these I would like to remember, in particular, F. Acerra, L. Bonolis, G. Celentano, F. Claro, A. De Gregorio, V. De Luise, A. Drago, E. Giannetto, C. Moreno, W. Mück, S. Pastor, B. Preziosi, G. Salesi, G. Senatore, P.D. Serpico, and the journalist R. Cavallaro.

Finally, I must thank the following for their collaboration: G. Paparo and L. Rossi from the Italian embassy in Buenos Aires, C.A. Barba from the *Museo Postal y Telegrafos—Museo de Telecomunicaciones* at the *Correo Central* in Buenos Aires, Anna Sicolo from the *Polo Archivistico Sanitario* of *Regione Campania* (Italy) and Edoardo Giaquinto and Giovanna Paola Perdonà, as well as the staff of several institutions: the Archives of the *Accademia delle Scienze detta dei XL* in Rome, the State Central Archives in Rome, the Archives of the University of Naples "Federico II", the Library of the Department of Mathematics "Guido Castelnuovo" of the Sapienza University in Rome, the "Giovanni Spadolini"

Library of the Senate of the Republic of Italy, the *Deutsches Museum* in Munich, *ETH* in Zurich, Niels Bohr Library, of the American Institute of Physics.

Most of the material appearing here was discovered or collected for the first time by Prof. Erasmo Recami and then published by him (as specified in this book). Such material is therefore copyrighted and it is used here with the kind permission of the copyright owner. It cannot be further reproduced without their written consent.

Contents

Prologue

When we leave the Central Station and head toward *Corso Umberto I*, then walk almost to the end, halfway between the *Quattro Palazzi*, a square bounded by four peculiar buildings, and *Piazza della Borsa*, we come to the historical home of the University of Naples, named after the sovereign who founded it, Frederick II of Hohenstaufen. Actually, this austere building has nothing of the Swabians (or *Svevi* as Italian's know them) about it, but it stands harmoniously among the other buildings along the avenue, which were restored between the end of the XIX century and the beginning of the XX. On the right-hand side of the university building—the other side looks onto the well-known *Via Mezzocannone*—we find the short *Via Antonio Tari*, and at number 3, just before climbing the three *Rampe del Salvatore* (the Saviour's Stairway), we find the front door of what used to be the Physics Institute.

This was the path which, on that day in January 1938, Ettore Majorana followed, just after being appointed full professor of theoretical physics "for his high repute of particular expertise [...] in the fields of studies concerning the said discipline". Then, for slightly less than three months, he passed through that doorway, before one day he decided to disappear. And he disappeared under mysterious circumstances, which remain to be fully clarified.

So why should we still be interested in Majorana, and in the short Neapolitan period of his life? Enrico Fermi, who knew him well, perhaps better than any other, explains it as follows.

> *I do not hesitate to declare, and I do not mean it as a hyperbole, that among all the Italian and foreign academics I have had the chance to meet, Majorana is the one, among all of them, that impressed me the most for the depth of his genius. Able to make inventive hypotheses and, at the same time, sharply criticise his own work and those of others, an expert calculator and a profound mathematician who never loses sight of the true essence of the physical problem under the veil of figures and algorithms, Ettore Majorana was endowed to the highest level with that rare set of aptitudes which make a first class theoretical physicist. And indeed, in the few years of his activity up to now, he has been*

able to bring his name to the attention of scientists all over the world, who have recognised in his work the sign of one of the best talents of our times and the promise of future achievements.[1]

Therefore, this is the story of those achievements, at least, as far Majorana wanted to take us.

Many of his relatives and friends witnessed the growing fame of the young prodigy from very early on. However, *"the sign of one of the best talents of our times"* is to be seen not in any surprising subjective evidence, but in his amazing scientific work, since his university years. His life itself, his choices, and his subsequent behaviour cannot be fully understood if one does not refer to it constantly. "He was not an eccentric: those aspects that could be interpreted as oddities were only the consequences of genius",[2] his sister Maria used to say. For those of us who did not meet him, all we can do, therefore, is try to understand Majorana the scientist, for *only then* will the image of the man take form more clearly. The present work starts from this premise. And sufficient new facts concerning the man of science have now come to light to give better shape to that image.

In our story Majorana is of course the main protagonist, but many other characters will prove to be of no less importance, people who were present from the beginning, and witnesses of Ettore's achievements, who have passed on their account to us. We shall therefore observe a peculiar dynamic in which secondary characters move from background to foreground, and not because they played any role in the protagonist's choices, but because his story has trickled down to us through these people, so that *their* choices may have determined (even unwittingly) our own perception of the events. The interest in what is presented here takes its origin directly from some of these characters, so that we are pulled into a swirling spiral that drags us ever further towards its centre.

The existence of the Moreno Paper with the Neapolitan lecture notes, the syllabuses of courses given as guest lecturer ("*libero docente*"), the discovery of a solution to an atomic model never found by others, to quote just a few, can certainly only have a limited character and impact. But they have also pointed to another direction worth investigating more deeply, and what was once obscure has now clearly revealed itself.

We have not obviously radically transformed what was already known, but we have tried to understand it better, and even those moments in Majorana's life and his personality traits, which seemed so clear in the words of witnesses, can now be seen in a new light. So can we say Majorana *ab omni naevo vindicatus*? Surely not, as there were hardly any results to compare with the new discoveries. Yet the emerging picture has even more charming and intriguing features. Anyway, in this work, we will purposefully avoid carrying out a psychological analysis of

---

[1]E. Fermi's letter to the Italian Prime Minister Benito Mussolini dated July 27, 1938, quoted in (Recami 1987). This document and similar ones disclosed by professor Recami and quoted hereafter are copyrighted and cannot be further reproduced without proper written permission.

[2]See, for example, the interview with Maria Majorana quoted in (Ferrieri and Magnano 1972).

Majorana. His acute sensitivity, the thoughts hidden deep in his soul, and even his state of health must certainly have weighed on the decisions he made, but here we can be quite sure that measurable facts guided our character's mind, and not vice versa. To quote Majorana himself, we will deal with "an incredible amount of things, but, as they are cheap facts of thought, and not empirical facts, one must just accept it".[3] Proceeding in this way, some lucky intuitions are necessarily going to be missed, but the reader can then usefully refer to the existing literature.

Our goal will be reached if only we are able to induce the feelings of Majorana's dearest friend, Giovannino Gentile:

> *Whenever he could see a stroke of genius, he was happy and became excited with enthusiastic admiration. Before his partner in studies, whose brilliance he was so proud of and whose disappearance just after they had begun university teaching he was to suffer painfully as the worst loss his studies could experience, he felt less than a disciple. He would often bring him to our house; and he wanted us, like him, to bow before that great genius and endorse the devoted humility with which he embraced him in such high respect.[4]*

But the time has come to raise the curtain and let Ettore Majorana step onto the stage.

Naples, Italy   Salvatore Esposito

---

[3]See the letter to Gastone Piqué dated October 17, 1927 in (Recami 1987).
[4]See (Gentile 1942). Thanks to E. Recami for providing a copy of the manuscript.

Part I
The Dostoyevskian Hero

Chapter 1
An Archimedes from Sicily Studies in Rome

From a distance he looked slender with a timid, almost hesitant, bearing; close to, one noticed his very black hair, dark skin, slightly hollow cheeks and extremely lively and sparkling eyes. Altogether he looked like a Saracen (Amaldi 1966).

Music, Books, and Science for a Fine Humourist

Ettore Majorana was born on 5 August 1906, on the second floor of an Art Nouveau building in *Via Etnea*, Catania. The following day, Giovannino Gentile, the one who was to become Ettore's dearest friend and colleague, was born in Naples, the son of an eminent philosopher.

Ettore "belonged to a branch of the family who were known on the island as 'the Archimedes of Sicily' because they had produced a long line of scientists, for a century or more without interruption" (Guarino 1950). His father Fabio Majorana, an engineer and the director of the local telephone company, was the youngest of the five sons of Salvatore Majorana Calatabiano, former Minister of Agriculture, Trade and Industry in two of Agostino Depretis' cabinets at the end of the previous century.

From a very young age, as the times required – I am talking about the years just before the Italian unification –, he spent his life planning a modern and free state, overcoming unspeakable difficulties and reaching the highest offices of the newly-united Italy from his hometown of Militello. Salvatore knew that a new state needed new men; hence his huge commitment to young people's education and, as a father, ever true to his ideas, he was uncompromising in his efforts to further the education of his sons: Giuseppe, Angelo, Quirino, Dante, and Fabio. He was duly successful there, as always. In the Majorana family everyone was free to pursue whatever suited them (Majorana 2007).

Among Ettore's uncles, Giuseppe was an economics and finance professor, and an esteemed author of several texts on these subjects; he was Dean of the University of Catania and a member of parliament in three legislatures. Angelo, professor of constitutional law and sociology when still a young man, was also Dean of the University of Catania, as well as being undersecretary and twice minister (for

© Springer International Publishing AG 2017
S. Esposito, *Ettore Majorana*, Springer Biographies,
DOI 10.1007/978-3-319-54319-2_1

Finance and Treasury) under Prime Minister Giovanni Giolitti. Quirino, on the other hand, was a famous and gifted experimental physicist, Augusto Righi's heir in Bologna, who ran the Italian Society of Physics for many years. Finally, Dante, a lawyer, was also a deputy and Dean of the University. Fabio Majorana's younger sisters, Elvira and Emilia, were educated in Rome and married respectively the State Councillor Oliviero Savini-Nicci and the lawyer Giovanni Dominedò. So the Majoranas could well be considered the pride of Catania (and of Militello in Val di Catania, their forefathers' hometown) during the XIX and XX centuries.

Ettore's mother, Salvatrice ("Dorina") Corso, belonged to a wealthy family of farmers from Passopisciaro, in the countryside around Mount Etna. Though "a strong woman, somewhat eccentric and very ambitious",[1] she had another side to her:

> Dorina was a tender mother, devoted to her family. I still remember the big parties she threw in her mansion to liven up her daughter Maria's evenings. The house was always full of friends. She was a careful and thoughtful mother to us as well, the estate manager's children. She treated us in the same way and she cared about our lives. I remember her on the day she received the news of Ettore's appointment for "special merit". She was walking up and down the patio, thoroughly excited, proud of her genius son.[2]

With her husband Fabio Majorana, before Ettore came on the scene, Dorina Corso had had several other children: Rosina (who later married Werner Schultze, a German soldier who had found shelter with the Majoranas during the war), Salvatore ("Turillo"), a graduate in law, and Luciano ("Luccio"), who became a civil engineer.

> Of Fabio's children, Salvatore, the eldest, devoted his life to the study of philosophy and the care of prisoners [...]. Luciano, my father, an engineer, devoted himself to extremely innovative projects, sometimes too innovative to be appreciated at the time; these ranged from electrical locomotives to car compressors. I still have a letter from Ferrari, the well-known car manufacturer, who, in 1966, felt that his device would be too difficult to apply to their complex engines... today many cars use the compressor. He also worked on optical devices and telescopes; Luciano signed several projects, including the big telescopes at Serra la Nave on Mount Etna and Campo Imperatore on Gran Sasso, still working perfectly today (Majorana 2007).

Years after Ettore, his little sister Maria was born. She later became a fine musician and piano teacher. Her recollections provide a vivid picture of Ettore's family environment:

> I have a very old childhood memory from our country home on Mount Etna, near Randazzo. In the evenings, the long summer evenings, [my father] used to read books out loud to his children, and it was something we all enjoyed a lot. His interests ranged from Dostoyevsky and Shakespeare to Goldoni. Nothing was left out. These reading sessions, performed by a father who was so nice to his children, will ever remain in my memories as something precious. [...]

---

[1]See the interview with Claudio Majorana in (Randazzo 1972).
[2]Interview with Peppino Cannavò in (Ventimiglia 2010).

My mother could play very well. She used to accompany my father, who had a good voice. She really loved opera. My mother had all of Puccini's and Bizet's scores. Ettore played too, by ear. All the brothers studied the piano. They were not particularly talented, but Ettore was a bit more so; he had an ear for it, and every now and then he would play something on the piano. [...]

He was very nice to me, very kind, and of course he helped me out with my maths homework. Once he even wrote an essay for me, a beautiful essay, which I later brought to school, very happy. Of course, it soon became clear that I was not the author. In that country house, in the summer evenings, when the air was clear and the stars shone bright in the sky, he would talk to me about stars, show me constellations, and name the stars, and I was happy.[3]

In their recollections, all Ettore's relatives note that he was a sensitive and loving person, often ironical, especially when referring to himself. "Ettore was a humourist"—his brother Luciano remembers—"He played around a lot, in his own way. For instance, he reproduced the old Sicilian dialect, impersonating the typical country folk of old, now no longer, and we all had great fun".[4]

Of course, his passion for science was inherited from his father Fabio and to an even greater extent from his uncle Quirino, with whom he kept in contact over the years (in person or from a distance, through exchange of letters) on various scientific matters. He also communicated regularly with his father (until his early death in 1934) and his brother Luciano, and mainly discussed philosophical and literary matters with his other brother Salvatore. All in all, he turned out to be the "Archimedes from Sicily" from an early age, and, as his sister Maria remembers, Ettore "was familiar with numbers even before he could speak. Our father would give him a packet of banknotes or a handful of coins and he would have fun adding them up".[5] Though stimulated in this way, however, it was soon clear that the young Majorana's mathematical skills were perfectly natural, and there are many stories to confirm this.

When he was five he could calculate the square and cube root of multi-digit numbers. For fun, he would crouch down under a table and, without pen or paper, carry out the most difficult calculations in his head using logarithms. I remember he must have been around six when, on the dock of the port, he was able to calculate to the hundredth part how much coal a given ship had to burn to follow a given route. "This child", the captain said, "will become a navy officer". But, at the age of nine, Ettore's destiny was already clear. He was unconditionally enthusiastic about scientific subjects. He did not talk much and avoided other people, but then suddenly, faced with an algebraic calculation or a physics topic, he would find his tongue again.[6]

His father Fabio personally took care of his son Ettore's early and marked mathematical abilities, together with his interest in physics and astronomy; he was actually tutored at home until the age of seven. But for the last elementary classes at

---

[3]Interview with Maria Majorana in (Bassoli 1998).
[4]See the interview with Luciano Majorana in (De Mauro 1965).
[5]See the interview with Maria Majorana in (Libonati 1966).
[6]Interview with Salvatore Majorana in (Libonati 1966).

the age of eight, his mother Dorina decided that the young prodigy should go to a boarding school in Rome, the prestigious Jesuit-run *Istituto Massimiliano Massimo*. Here Ettore met one of his few real friends, Gastone Piqué.

> He was there at the Massimo – fourth or fifth grade, I cannot remember – walking down the corridor. Some students had just come out of their finals, where they had been doing the maths test. They were holding their papers and saying how difficult the test was, and how no one had understood a thing. And along he comes, looks briefly at the paper, then stands in a corner and after a couple of minutes has solved the problem. Without writing it down. In his head. [...]
>
> As he was not physically good-looking, he had a complex about women [...]. At high school he had a girlfriend. She was a prefect's daughter. She was a very bright woman. She was charmed by this prodigy, this young genius. But for his part, he felt nothing, in fact he rejected her, because he must actually have had an inferiority complex about his ugliness.[7]

Lured by the capital city, his mother and the other children then moved to Rome in 1921. Three years later their father Fabio joined them, when the telephone company he was working for was absorbed by the government and he was relocated. Then, following the common practice, for his last year in high school, Ettore (and his brother Luciano) moved to the equally prestigious *Liceo Statale "Torquato Tasso"* in Rome, where he got his *diploma* in the summer session of 1923.

An Engineer's Career

Ettore Majorana enrolled in the *Biennio* (first two-year period) of the engineering course at the University of Rome on 3 November 1923 (although the actual matriculation was on the following 5 December), and he regularly attended lectures and practical sessions.

During the first-year, students were expected to attend 5 courses (plus practical sessions), followed by 3 exams; in the second year there were 5 courses and as many exams. Here are the details for the student Majorana:

1st year (1923–4)

Course (professor)	Exam (mark)
Algebra (Severi) and practical (Tricomi)	24/06/1924 (30/30)
Analytic and projective geometry (Castelnuovo) and practical (Lucaroni)	16/06/1924 (30 *cum laude*)
General chemistry (Parravano)	04/11/1924 (27/30)
Physics (Collodi)	
Drawing (Armanni)	

---

[7]Interview with Gastone Piqué in (Fiori 1971).

2nd year (1924–5)

Course (professor)	Exam (mark)
Rational mechanics (Levi-Civita) and practical (Bisconcini)	09/07/1925 (30 *cum laude*)
Probability (Severi) and practical (Bilancini)	29/06/1925 (27/30)
Physics (Corbino)	02/07/1925 (30/30)
Descriptive geometry (Pittarelli) and practical (Loudadari)	29/06/1925 (30/30)
Drawing (Armanni)	15/06/1925 (18/30)

If we look at his marks, it is no accident that the young Majorana was strongest in those courses that interested him the most, and also those that were given by the best experts in their respective fields. Indeed, his first exam, obtained with honours, concerned the course on analytic and projective geometry given by Guido Castelnuovo. The latter was born in 1865 in Venice and, together with Corrado Segre, Federico Enriques, and Francesco Severi, founded the Italian school of algebraic geometry, which, for a long period, enjoyed high international fame. Castelnuovo, however, was not only a gifted mathematical researcher (active until the beginning of the XX century), but also an appreciated teacher (he had held the tenure at the University of Rome since 1891), when Majorana met him. One of Ettore's university colleagues, Emilio Segrè, remembers that "Castelnuovo was a paragon of clarity, and in spite of his somewhat soporific voice, one learned new and interesting things" (Segrè 1993). The peculiar feature of his course, which aimed mainly to stress the fundamental ideas and methods underlying the given subject, can be most clearly expressed in his own words: "every question is discussed here using the most suitable method to analyse it, and several topics, examined from different angles, acquire a unique significance".[8] We cannot help but note how Castelnuovo's goal would have a deep impact on the development of the young Majorana's personality, and as we shall see, the following observation is quite appropriate for the future professor at the University of Naples, provided of course that we change the specific interests: "I am confident that [the course] will achieve the double goal of providing engineering students with the fundamental geometrical notions, without boring them with useless oddities, and encouraging those focused on science to broaden their own culture in higher fields."[9]

Among Ettore's other outstanding professors, distinguished mathematical personalities of the day who did not fail to notice their student's peculiar gifts, we find Tullio Levi-Civita and Francesco Severi. The former (born in 1873 in Padua) was a truly exceptional mathematician with a remarkable geometric intuition which he applied to a variety of different physical problems. He is remembered for developing abstract differential geometry with Gregorio Ricci-Curbastro. This is the mathematical formalism required by Einstein's theory of general relativity. He was

---

[8]Preface to the first edition of (Castelnuovo 1956).
[9]Preface to the second edition of (Castelnuovo 1956).

granted the tenure of rational mechanics at the University of Rome in 1919, after achieving the same at the University of Padua in 1897, when he was only 24. One can see how deeply he dealt with his course in the classic three-volume essay written with Ugo Amaldi (Levi-Civita and Amaldi 1923). This has marked the education of generations of university students. "He was one of the most eminent professors in Italy for over 40 years and attracted students from all over the world, whom he encouraged with his patience and nobility. Kindness and modesty were manifestations of his soul, and many benefited from this kindness, retaining a lasting memory of his extraordinary personality" (Nastasi and Tazzioli 2005). Direct evidence of this is given by Segrè himself.

> Levi-Civita's course on rational mechanics was poorly attended, although the professor was famous and the lectures were good, even if slightly verbose. Levi-Civita was very short and also short-sighted; nevertheless, he strove to reach the top of the blackboard, putting his nose very close to it, raising his arm, and writing blind. In this position, he was once struck on the back of the head by a missile from the peashooter of some nasty student. Levi-Civita turned around and, with the most innocent expression, asked: "Have I got the sign wrong?" His candor and good faith were so obvious that nobody laughed, and no peashooter ever dared disturb him again (Segrè 1993).

Francesco Severi (born in 1879 in Arezzo) arrived in Rome in 1922, just a few years before Majorana matriculated. Like his colleagues mentioned above, he was a very skilled teacher and researcher in the field of analytic and projective geometry. Segrè remembers: "Severi gave excellent lectures, and I was pleased by the change of level from high school; here there was a real intellectual stimulus and a challenge to understand, perhaps even to try to invent something new" (Segrè 1993). He was known for his rather harsh personality: "Personal relationships with Severi, however complicated in appearance, were always reducible to two basically simple situations: either he had just taken offence or else he was in the process of giving it" (Roth 1963). An anecdote concerning Majorana is of interest here, as reported by Segrè:

> Once, not having sufficiently prepared a lecture, Severi started a proof of a theorem the wrong way. Majorana immediately whispered that he would soon be in trouble, so we all anticipated what was to come. After a minute or two, Severi's face reddened, and it became obvious that he did not know how to proceed. Some voices then murmured: "Majorana predicted it." Severi did not know who Majorana was, but said haughtily, "Then let Mr. Majorana come forward." Ettore was pushed to the blackboard, where he erased what Severi had written and gave the correct proof. It is noteworthy that Severi neither complimented him in any way nor, as far as I know, made any effort to become acquainted with him (Segrè 1993).

Another anecdote, once again reported by his friend Segrè, is about the old professor of descriptive geometry, Giulio Pittarelli, who was born in 1852 in Campochiaro in the region of Molise, but graduated in Naples in mathematics in 1874 and two years later in engineering.

> While I was waiting to be called to an oral examination, Majorana gave me a synthetic proof for the existence of Villarceau's circles on a torus. I did not fully understand it, but memorized it on the spot. As I entered the examination room, Professor Pittarelli asked me,

as was his wont, whether I had prepared a special topic. "Yes, on Villarceau's circles," I said, and I proceeded immediately to repeat Majorana's words before I forgot them. The professor was impressed and congratulated me on such an elegant proof, which was new to him (Segrè 1993).

A non-negligible part of the mathematical courses was devoted to practical sessions, held regularly by the assistants[10] to the tenured professors. These assistant professors "were generally more accessible than the others. Severi first had Tricomi[11] and then Fantappiè,[12] who proposed very interesting sessions. [...] Castelnuovo's assistant,[13] on the other hand, was expert but did not broaden our horizons" (Segrè 1993).

While mathematics courses constituted the lion's share of the first two years at university, other less "theoretical" but fundamental courses had nevertheless to be attended by the students, such as those on general chemistry and experimental physics.

The course on chemistry was given by Nicola Parravano (born in 1883 in Fontana Liri), a well-known metallurgical chemist who had trained in Berlin, and whose importance in the Italian chemistry community was growing at the time,

---

[10]The practical sessions of Levi-Civita's course were held by Giulio Bisconcini (born in 1880), who was appointed assistant of analytical and projective geometry at the University of Rome through the recommendation of Levi-Civita himself to Vito Volterra. Though a lecturer of rational mechanics, he nevertheless taught in high schools and became a mathematics teacher at the *Istituto Commerciale "Luigi di Savoia—Duca degli Abruzzi"* in Rome.

[11]Francesco Giacomo Tricomi (born in Naples in 1897) was first an assistant at the University of Bologna, and then moved to Rome university (which he eventually left when he accepted the tenure at the universities of Florence and later Turin). Already in 1923 he published one of his most famous papers, which contained the study of the differential equation named after him. The equation turned out to be relevant to the description of an object moving at supersonic speed. Guido Castelnuovo's daughter Emma made the following revealing remarks about his personality: "he had a difficult character and was known for that", but "I must say he helped me a lot" in escaping from racial persecutions (in *La fantasia e la memoria. Conversazione con Emma Castelnuovo*, edited by R. Natalini and M. Mattaliano, http://matematica.uni-bocconi.it/castelnuovo/castelnuovo.htm).

[12]Luigi Fantappiè was born in Viterbo in 1901 and graduated in Pisa in 1922. He was Severi's assistant at the University of Rome from 1925 to 1927, when he was granted the tenure of calculus first in Cagliari, then in Palermo, and eventually in Bologna. From 1934 to 1940 he was in São Paulo in Brazil to found the Institute of Mathematics at the university there; from 1940 on he was offered the tenure of advanced calculus at the *Istituto Nazionale di Alta Matematica* (Institute of Advanced Mathematics) of the University of Rome, founded and directed by Severi. Severi's assistant for the course on probability was Raul Bilancini, for whom no biographical account appears to be available.

[13]Raffaele Lucaroni (born in Ancona in 1887) was the assistant for several courses at the University of Rome, "a very effective teacher, well known around Rome", in G. Castelnuovo's words, who asked him to become professor at the clandestine university for Jewish students in Rome. Because of his refusal to take fascist membership, he was forced to teach only in private institutions from that moment on. He was the only mathematician, together with Castelnuovo and Enriques, to attend Volterra's funeral (one of the few university professors who refused to sign the oath of allegiance to the fascist regime) in October 1940 (Della Seta 1996).

until he eventually became one of its leading figures at the end of the 1920s. After getting tenures in Padua, Florence, and Milan (where he founded and directed the Breda Scientific and Technological Institute in Sesto S. Giovanni), he moved to the Roman athenaeum, where he was also appointed head of the Faculty of Science. A committed supporter of fascism, he advocated a certain idea of science as a "social force", and considered the "fascist scientist" to be "a man of culture, a technical innovator, and a political and ethical individual" (Parravano 1936). He held many positions of responsibility for the fascist regime, and also became a member of the Academy of Italy. His characteristics as a teacher are well summed up by Segrè:

> I soon came to suspect, however, that the chemist, Parravano, did not always know what he was talking about. At home I had found a treatise on physical chemistry by Walther Nernst, and comparing what I had learned from it with what Professor Parravano taught, I concluded that he had misunderstood several things, or at least that he had understood them differently from me (Segrè 1993).

The regular teacher of the course in experimental physics was Orso Mario Corbino (born in 1876 in Augusta), an important personality of his time, both scientifically and institutionally. In 1898, two years after graduation, together with his teacher Damiano Macaluso, he discovered a remarkable property of light known as the *Macaluso-Corbino effect*, and in 1904 he won two university selections at the same time, accepting the tenure of experimental physics at the University of Messina. Four years later he moved to the University of Rome where, in 1918, he became director of the Institute of Physics, in *Via Panisperna*, a position he kept until his death in 1937. Between 1918 and 1922, he discovered a variant of the Hall effect, and later, with the collaboration of his assistant Giulio Cesare Trabacchi, he devised an electrotechnical tool able to produce high voltages to feed X-ray generators for use in radio-diagnostics and radiotherapy.

However, since his move to Rome, he was involved not only in science, but also in public activities. He was elected senator of the Kingdom of Italy in 1920, and the following year became Minister of Education in the Ivanoe Bonomi cabinet, before going on to become Minister of National Economics in 1923–4, directly appointed by Mussolini, even though he was not a member of the fascist party. As we shall see later, in the following years he directed his efforts to setting up a school of modern physics in Rome (the first and only one in Italy until the post-war period), one which could hold its own against the world's greatest physics centres, a task achieved mainly by Enrico Fermi and Franco Rasetti. Corbino, however, did not restrict himself to ensuring his young protégés the relevant institutional support; he followed their research with great personal involvement and care:

> Simultaneously with his scientific activity, Corbino carried out his teaching activity. He always loved being surrounded by young people, giving them a lot of detailed advice based on his experience and knowledge. [...] He loved discussing their experiments with them, both the plan and the actual, always picking up the essential points and avoiding unnecessary details. [...] An exceptionally brilliant and witty speaker, he knew how to liven up his lessons and talks, so that they were pleasant to attend, and made the most difficult topics easily understandable (Fermi 1937).

He was thus a great teacher, but he followed the prevailing fashion of the time in Italy of teaching only topics in classical physics, as Segrè himself remembers: "He taught electromagnetism clearly and efficiently, but at an elementary level. In the first two years no one taught thermodynamics or optics, not to mention any modern topics" (Segrè 1993).

In the first two years of engineering studies, there was a course on drawing (followed by an exam at the end of the second year). One such course was given by the architect Osvaldo Armanni, professor at the University of Rome from 1907 to 1929, known as the designer (with Vincenzo Costa) of the *Tempio Maggiore* (the Great Synagogue) of the Jewish community in the capital (inaugurated in 1904). Ettore was probably not much interested in the kind of topics dealt with there, and as a matter of fact he passed Armanni's exam (the first of his second year) with the lowest mark.

Regia Scuola di Applicazione

During his studies in engineering, Ettore would spend his free time with some friends, including Emilio Segrè, a brilliant young man from a well-to-do family in Tivoli. In fact, when not studying, he spent much of his leisure time with his brother Luciano, who also enrolled at the *Biennio di Studi di Ingegneria* (preparatory biennium) when he and Ettore graduated from high school. A dearest friend, dating back to his school days, was Gastone Piqué, who joined the Majorana brothers in the same studies. Other friends were Enrico Volterra, son of the well-known mathematician Vito, Giovanni Enriques, also the son of a famous mathematician, Federigo, and Giovanni Ferro-Luzzi.

Once the preparatory two-year course was finished, the next step was to enrol at the *Scuola di Applicazione degli Ingegneri* (Engineering School) for another three years of studies, which Majorana started on 3 December 1925.

> In my third year at university, I transferred to the Engineering School, where I found the courses much less interesting than in the preparatory biennium, except for one by Professor U. Bordoni, who taught us thermodynamics according to Clausius, emphasizing all its subtleties. The other professors taught ordinary engineering practice, at a low technical level and without imagination (Segrè 1993).

Ettore was also to notice the new environment, and began to show signs of impatience.

> He thought we were lingering too long over the description of inessential details, while not enough emphasis was given to general synthesis, important for a sound scientific framework. This deeply-rooted belief of his was the origin of frequent lively and sometimes heated arguments with some of the professors (Amaldi 1968).

And indeed, as further proof of this, one only has to look at his examination results (see later on), some of which were certainly not brilliant! Once Ettore even roundly failed hydraulics, but a close friend explained what happened this way:

One thing is sure: Ettore attended engineering until the fourth year without buying a single book. He was not keen on engineering, always engaged in his personal studies. Two days before the exam, he would borrow the books from me and quickly browse them; that is the reason why once he even failed.[14]

Anyway, Majorana never failed to help friends in trouble, something he had already done in the past, as we have seen (and which also went on, in different ways and for different purposes, after graduation). Gastone Piquè remembers:

On the day before the theoretical physics exam I could not figure out a problem. I no longer remember which one, but it filled about four pages of text. So I turned to Ettore and asked for his help. And he told me: "You see, those four pages might as well be summed up in four words", and gave me a crystal clear explanation. By a happy coincidence that happened to be the content of my oral exam. And the professor was astonished: how had I been able to find that explanation?[15]

And, even more ironically:

Once, when I was an engineering student, in order to do a favour to a friend, he showed up at an exam in his place, thereby getting him a very good mark.[16]

The third and fourth year university courses (the first two years at the Engineering School) that Ettore attended were as follows, with the corresponding exam results:

3rd year (1925–6)

Course	Exam (mark)
Applications of descriptive geometry	05/11/1926 (70/100)
Applied chemistry	24/11/1926 (95/100)
Elements of production	
Technical physics	12/06/1926 (100/100)
Geodesy and topography	14/07/1926 (75/100)
Static graphics	31/07/1926 (85/100)
Applied hygiene	25/06/1926 (100/100)

4th year (1926–7)

Course	Exam (mark)
General architecture	
General electrotechnics	21/06/1927 (100/100)
Hydraulics	02/12/1927 (75/100)

(continued)

---

[14]Interview with Gastone Piqué in (Castellani 1974).

[15]Interview with Gastone Piqué, *loc. cit.*

[16]Interview with Maria Majorana in (Castellani 1974).

(continued)

Course	Exam (mark)
Law	16/07/1927 (100/100)
Mechanics for construction	29/07/1927 (90/100)
Mechanics for machinery	17/11/1927 (85/100)
Mineralogy and geology	04/07/1927 (100/100)

In these two years some of the purely engineering courses regularly attended by Majorana were not followed by exams. The promising career of our aspiring engineer was about to take a dramatic turn.

Corbino's Appeal and the School of Physics in Rome

In June 1927, the director of the Institute of Physics at the University of Rome, Orso Mario Corbino, who also ran the course in experimental physics for engineering students, made an odd request. It was a time particularly rich in important new discoveries for physics in Europe, and according to professor Corbino, the fact that Enrico Fermi was being asked to accept a tenure at the University in Rome could open up exceptional new prospects to young men gifted and willing enough to devote themselves to research in physics.

He actually meant to "recruit" some of the better elements who, once trained in the study of physics, would go on to make some valuable contribution to the small research group just built up around Fermi.

Enrico Fermi was born in Rome on 29 September 1901, and attended the *Scuola Normale Superiore* and the University of Pisa, where he graduated in 1922. "What was remarkable about his studies was the way he was able to choose the most relevant subjects, never wasting his time on details, no matter how intriguing they were" (Segrè 1960). During his lifetime he met a lot of outstanding scientific personalities, first as a young researcher during his trips to Göttingen with Max Born and Leiden with Paul Ehrenfest, then as a prominent scientist both in Rome during the 1930s and in the United States after he received the Nobel Prize for Physics in 1938. Among the many who knew him, Hans A. Bethe described his role on the international scientific scene:

> Fermi was unique among the great physicists of the twentieth century in being one of the greatest in experimental physics and at the same time being one of the greatest theoretical physicists. He was unique also in the width of his contributions. He may have been one of the last physicists who knew almost all of physics and used it in his research. [...] Wherever new and important frontiers were opened in physics, Fermi was there to lead the way and to show us how to make progress (Bethe 1955).

Such opinion may seem somewhat exaggerated, but is actually well testified by his scientific work (Fermi 1960), and it was always shared by anyone who had ever met him and appreciated his skills since he was a student. Even Luigi Puccianti, the professor of experimental physics and director of the Institute of Physics in Pisa, recognised his exceptional skills: "Puccianti very soon saw that he had little to teach but much to learn from his student Fermi. He acknowledged this with the utmost candor and many times asked his student to 'teach me something'" (Segrè 1960).

A peculiar trait of his scientific personality that would remain with him throughout his life was put like this by Segrè: "[...] although never repulsed or frightened by any mathematical difficulty, [he] is not interested in mathematics for its own sake. Whether a theory is easy or difficult does not seem to concern him; the important point is whether or not it illuminates the essential physical content of the situation. If the theory is easy, so much the better, but if difficult mathematics is necessary, he is quickly resigned to it".[17]

But the turning point in Fermi's life (and in the Italian physics of the day) came after graduating, when he "went back to his family in Rome and introduced himself to the Senator Professor Orso Mario Corbino, director of the Institute of Physics at the University. From this encounter, of the utmost importance for his life, a lively and mutual fondness developed, and high respect too, which went on until Corbino's untimely death. Corbino soon understood the young *Normale* student's genius, and thought he was the right person to help physics flourish again in Italy, supporting one of his greatest aspirations. From that moment on, Corbino did whatever possible to facilitate the young man's career and help him found a school" (Segrè 1960). The first decisive step was to advertise for the theoretical physics tenure—for the first time in Italy—at the University of Rome. The ministerial committee gathered to examine the candidates who had applied for the post. Their members were A. Garbasso, G.A. Maggi, M. Cantone, Q. Majorana, and O.M. Corbino, who did not hesitate to give Fermi the post on 9 November 1926:

> After examining professor Fermi's vast and complex scientific work, the Committee is unanimous in recognizing his exceptional qualities and observes that, despite his young age and few years of scientific work, he is already bringing great honour to Italian physics. He totally masters the subtlest resources of mathematics and is able to use them efficiently and parsimoniously, never losing sight of the physics problem he is trying to solve and the value of the physical quantities he has to deal with. While he is perfectly familiar with the most intricate concepts of mechanics and classical mathematical physics, he is fully proficient in the most difficult problems of modern theoretical physics. For these reasons he is today the most skilled and the worthiest representative of our country in this time of intense scientific activity around the world. The Committee is therefore unanimous in declaring professor Fermi suitable for the tenure of theoretical physics, which was the aim of this selection process, and believes he represents the best hope for the promotion and future development of theoretical physics in Italy.[18]

---

[17]See the biographical note by Segrè in (Fermi 1960, p. XVII).

[18]Report of the Examination Committee, in *Bollettino del Ministero dell'Educazione Nazionale* (Part II: Administrative Acts), Year 54 (1927), p. 634.

Once in the post, Fermi "soon felt the need to bring together a group of students and future collaborators. Moreover, in his new idea of physics, theory and experiment were inseparable; that is why he wanted a young experimental colleague, F. Rasetti, with whom he had been working since his time in Pisa. So he introduced Rasetti to Corbino, who later hired him as assistant" (Segrè 1960).

Franco Rasetti, born in 1901 in Pozzuolo Umbro, chose to study engineering at the University of Pisa, where he met Fermi, a first year physics student.

> After the courses had begun in the autumn of 1918, I met Enrico Fermi, also a first year student (of physics) who was attending the same classes. I immediately had the impression of an extraordinary person, due to his mature outlook and exceptional knowledge and understanding of mathematics and physics. [...] It was undoubtedly through Fermi's influence that, at the end of the second year, I decided to leave engineering and become a student of physics.[19]

At the University of Pisa, he graduated in 1922 with a thesis on spectroscopy, under the supervision of professor Luigi Puccianti, who had already opened the doors of his laboratory to the two friends (and Nello Carrara). They started working closely together, later moving to the University of Florence, where Rasetti became Garbasso's assistant. When Fermi was called for the tenure in Rome, he and Corbino realised that they would need a good experimental physicist in the Rome group. Rasetti was their choice (Corbino had already met him in 1925), and so he became Corbino's assistant in 1927. He fitted perfectly in the growing Rome group and became the experimental physicist par excellence, or as Nicola Cabibbo put it, "he was the *armed wing* of Enrico Fermi's group; [...] he would even call him back from his holidays in Morocco to confront him with the planning and realization of a delicate and sophisticated experiment".[20] In the following years Rasetti visited important laboratories abroad, and in particular in 1928–9 he went to the United States to meet R. Millikan, at the California Institute of Technology in Pasadena. There, on his own initiative, he began some studies on the Raman effect in gases which turned out to be fundamental to understand the composition of atomic nuclei. In 1930 he was granted tenure at the University of Catania, but Corbino, who did not want to lose this valuable physicist, had the tenure of spectroscopy expressly created at the University of Rome, and it was duly offered to Rasetti. As Corbino himself wrote:

> The gifts of a brilliant and careful experimenter and a deep knowledge of theoretical physics achieve a rare balance in Rasetti; his knowledge extends to the most recent developments and he is also a first-rank planner. His ability is universally acknowledged both in Italy and abroad, where he has often been invited to give lectures.[21]

---

[19]F. Rasetti, unpublished autobiographical note, kept in the Amaldi Archive at the Department of Physics, Sapienza University, Rome.

[20]N. Cabibbo, article in *Il Messaggero*, December 7, 2001.

[21]O.M. Corbino, unpublished note in the Amaldi Archive at the Department of Physics, Sapienza University, Rome.

The place where Fermi's developing group carried out their research was the Institute of Physics at the University of Rome, next door to the chemistry department, at 89a *Via Panisperna*.

> The physics building was perfectly adequate for scientific work in the 1920s, and it compared favorably with other major European laboratories. The equipment, however, was somewhat disappointing, including mainly spectroscopes and optical instruments, with only a few good modern pieces of apparatus. The workshop was old-fashioned as regards the machine tools and not up to the task, but the library was excellent and fully up to date. [...]

> The third floor of the building was Corbino's residence. The second floor contained the research laboratories and the offices of Corbino, Lo Surdo, and Fermi, as well as the library. The first floor contained the workshop, the classrooms, and the students' laboratories. The basement contained the electric generators and other facilities.

> Fermi, Rasetti, and their students occupied the whole south side of the second floor; Lo Surdo most of its north side. [...] Neighboring quarters were occupied by Professor G.C. Trabacchi, who was the chief physicist at the Health Department (*Sanità Pubblica*). He had an excellent supply of instruments and materials which he generously lent whenever we needed them (Segrè 1970).

The original nucleus of the rising Rome group was thus constituted and the plea to the students of engineering brought good results. Edoardo Amaldi accepted Corbino's invitation and in June 1927 moved to physics. Later that same summer Emilio Segrè, who had met Rasetti at the University of Florence, and through him also Fermi, decided to stop their studies at the Engineering School and enrol in physics.

Fermi Passes an Exam

And what about Majorana? After several attempts, his university friend Emilio Segrè, aware of the fact that an interview with Fermi would help the young Ettore to discover his peculiar aptitude for research in physics, finally succeeded in convincing him to meet Fermi. Segrè himself arranged the meeting, after proclaiming Majorana's exceptional abilities at the Institute of Physics in *Via Panisperna*.

> You could see [his brilliance] because you could ask Majorana things that you couldn't ask anybody else in the world. He was a prodigy. At least, I mean, one couldn't know whether Majorana would become a second Newton, but you could get him to integrate, I don't know, a very complicated integral and so on; he would look at it and tell you the answer without even writing it down. He could do feats of this type, plus numerical feats, but the numerical feats were more of a kind of showmanship...[22]

So one autumn day in 1927 Ettore went to see Fermi who, without ceremony, described the research they were carrying out at the time.

---

[22]T.S. Kuhn's interview with E. Segrè, May 18, 1964. Niels Bohr Library and Archives, American Institute of Physics, https://www.aip.org/history-programs/niels-bohr-library/oral-histories.

For almost a year then Fermi had been working on the physics of heavy atoms, which would thus carry many electrons. He introduced some statistical concepts whose consequences he himself and others before him had studied and which had led to the well-known *Fermi-Dirac statistics*. Application of these concepts to atomic physics led to the formulation of the statistical model of atoms known as the Thomas-Fermi model. The characterization of heavy atoms in this model depends on knowing a mathematical function which is the solution to a complicated equation (in mathematical terms, it is a nonlinear differential equation with particular boundary conditions). This is the "Fermi universal potential", valid for any chemical element. Unfortunately, the solution to this mathematical problem had not yet been found, but Fermi, who was certainly not the kind of person to be dissuaded by mathematical difficulty, had found an approximate solution to the equation and had drawn up a table showing its numerical values. This table was widely used by the atomic physics community. For example, one of the fathers of quantum theory, Arnold Sommerfeld, while developing a different approximation to the Thomas-Fermi function, checked the validity of his approximate solution by comparing it with the values in Fermi's table.

That day, in Fermi's office, Majorana listened carefully to what Fermi was saying, asking for several clarifications, then left the Institute. The witnesses to that meeting, Rasetti, Segrè, and Amaldi, give us the following recollection:

> The next day, towards the end of the morning, [Majorana] came back to the Institute, went straight into Fermi's office and asked him without further ado to show him the table which he had seen for a few moments the day before. Once in his hands, he took from his pocket a small piece of paper on which he had worked out a similar table at home during the last 24 hours. He compared the two tables and, seeing that they totally agreed, said that Fermi's table was right, went out of the office, and left the Institute (Amaldi 1966, 1968).

What might seem nothing more than an amusing anecdote, which already perfectly illustrates certain aspects of Majorana's personality, has in recent years undergone a careful scientific check (Esposito 2002a; Di Grezia and Esposito 2004). In one of Majorana's personal notebooks, those known as *Volumetti*, which are kept in Pisa, we find that night's output written up in a few pages. Majorana explored *three* possible ways to solve the Thomas-Fermi equation and, after giving the problem a firm mathematical formulation, he obtained the series solution with an original method which can even be applied to a whole class of mathematical problems (Esposito 2002b), obtaining the above-mentioned numerical table. Unfortunately, this work remained unknown in its details (even to those who witnessed the meeting) until a few years ago, and it is worth noting that, while some of Majorana's conclusions long anticipate the results of several well-known authors (both mathematical and physical), other results have never been obtained by anyone else.

To continue with Amaldi's story, "some days later [Majorana] switched to physics and began to attend the institute regularly" (Amaldi 1968). Fermi had therefore passed the test... and Majorana felt ready to move from his studies in engineering to more interesting ones in physics.

Graduation in Physics

His admission to the fourth year of the degree course in physics is officially registered on 19 November 1928, but there is solid evidence that Ettore actually
changed to physics about a year earlier, by the end of November 1927, or at least in
the first few days of 1928, as Amaldi (1968) remembers, that is, at the beginning of
his fifth year at university. One piece of evidence comes, for example, if one
compares Fermi's papers on the statistical model of atoms (the Thomas-Fermi
model) with the calculations found in Majorana's *Volumetti*.

Though Majorana had already distinguished himself among his peers for his
abilities at the Engineering School, his personality seems to emerge even more
clearly in this new physics environment. "Everyone was soon struck by Ettore
Majorana's lively intellect, depth of understanding, and range of studies, which
made him stand out from all his new colleagues" (Amaldi 1968). This is reflected in
the results of the exams given at the Physics Institute and listed here:

4th year physics (1927–8 and 1928–9)

Course (professor)	Exam (mark)
Theoretical physics (Fermi)	05/07/1928 (100 *cum laude*)
Higher physics (Lo Surdo)	02/07/1929 (30 *cum laude*)
Mathematical physics (Volterra)	21/06/1929 (30 *cum laude*)
Earth physics (Fermi)	27/06/1929 (30 *cum laude*)
Practical physics (Rasetti, Segrè)	05/07/1929 (30/30)

Majorana attended the course in theoretical physics when he was still a year 2
student at the Engineering School (year 4 at the university), a fact reflected by his
mark in the related exam. Furthermore, other courses, such as Rasetti's practical
physics and Lo Surdo's higher physics, were "optional" for students of engineering,
while they were "compulsory" for physics students. Among the compulsory courses
there was a course on chemical preparations; but, after a decree of 15 June 1929, the
Faculty of Science validated the exam in applied chemistry as a substitute, and
Majorana had already passed this. Therefore, the exams Ettore sat to get his degree
in physics are the following, as attested by a certificate of studies released on 15
May 1964, upon Amaldi's request (Amaldi 1968):

1. Algebra, 30/30
2. Analytic and projective geometry, 30 *cum laude*
3. Applied chemistry, 27/30
4. Rational mechanics, 30 *cum laude*
5. Drawing practice with elements of machinery, 18/30
6. Probability, 27/30
7. Descriptive geometry, 30/30
8. Physics, 30/30

9. Higher physics, 30 *cum laude*
10. Earth physics, 30 *cum laude*
11. Practical physics, 30/30
12. Mathematical physics, 30 *cum laude*.

Notice how strange it is that, among these exams, there is no mention of the theoretical physics course with Fermi, probably because it was considered "optional" in the engineering curriculum.

A general overview of Majorana's career as a university student brings out not only his peculiar mathematical abilities, but even more so his grasp of physics. We may at times have lingered too long on the idea of his being a "mathematical genius", and hence a sort of "attraction", rather than considering him as an all-round scientific personality, something which is crystal clear if we look at his personal notebooks, although they were absolutely unknown at the time (Esposito et al. 2003). In any case, it should also be said that not many people would then have been able to fully appreciate such extraordinary abilities, in contrast to certain outstanding physicists such as Fermi, Levi-Civita, and Castelnuovo, as already mentioned, but also Lo Surdo and Volterra.

Antonino Lo Surdo was born in Syracuse, Sicily, in 1880. He was the one who, in 1913, independently from J. Stark, discovered the effect that an electric field has on the emission of light (*Stark-Lo Surdo effect*). Since 1919, he had held the tenure of advanced physics at the University of Rome, where he carried out his experimental research, mainly on spectroscopy. However, since 1908, when he lost his entire family with the exception of his brother in the catastrophic Messina earthquake, he had also been working on seismology and geophysics, and in 1936, backed up by Guglielmo Marconi's C.N.R., he founded the National Institute of Geophysics to "promote, execute and coordinate the study of Earth's physical phenomena and their practical applications". It is interesting to see what two of his students say about his activities as professor of advanced physics, as Majorana knew him. Segrè says that his lectures at university "were essentially expositions of Drude's books on optics (1900) and J.J. Thomson's work on gas discharges (1903); they made no significant mention of quantum theory" (Segrè 1970). Another pupil, Mario Ageno, who attended the course some years later, remembers that Lo Surdo

> was not considered to be, and probably was not, an outstanding physicist. But he was actually the only one to take serious care of the students, and his course was an ongoing series of wonderful experiments, through which he was able to highlight the most deeply hidden phenomena. And it was there that I realized that physics is not a subject to be studied in books, but a science of nature.[23]

So, although Ettore was not at all keen on the experimental side of things, we may reasonably guess that he took an interest in Lo Surdo's course anyway,[24] as is

---

[23]Interview with Mario Ageno in *Sapere*, vol. 59, no. 4 (1993).

[24]It is interesting to note that, among all the manuscripts of personal notes left by Majorana, and which deal with theoretical questions, only in one of his *Quaderni* is there a reference to experimental questions (not research), referring to topics in the course on advanced physics.

clear from his excellent exam result; all the more so considering Lo Surdo's well-advertised hostility towards Fermi and those in his group.

A totally different case was the course in mathematical physics given by the well-known mathematician Vito Volterra (born in Ancona in 1860). He graduated in physics in 1882, at the University of Pisa, where he also attended the *Scuola Normale Superiore*; the following year, when he was only 23, he became professor of rational mechanics in the same athenaeum. His name is universally known for his studies on functional analysis (which he founded) and the related integral and integro-differential equations named after him. In 1900, after being solicited by several universities, he was granted the tenure in mathematical physics at the University of Rome, where he started to work on various practical applications of his mathematical studies. In the meantime he was a volunteer in the First World War, working on airships and air balloons, and had the idea of filling them with inert helium rather than flammable hydrogen. In 1905 he was named Senator of the Kingdom of Italy and, with the advent of fascism, he became a strong opponent in Parliament. In 1931 he was one of the few university professors who refused to sign the oath of allegiance to the regime, and as a consequence he was forced to resign his university post. Concerning the course Majorana attended, his friend Segrè recalls:

> Volterra's lectures were well organized and the subject matter was skilfully chosen (as I realized later), but his delivery, in a thin and slightly high-pitched or nasal voice, tended to send me to sleep. There were no textbooks, so one had to take notes; I therefore asked Amaldi to write for both of us, since he wrote faster than I, and also to wake me up if I fell asleep. Volterra used to close his eyes while lecturing and somebody said that this was because, being kind-hearted, he did not want to see the students' sufferings. Except for these superficial shortcomings, the lectures were profitable. One learned the mysteries of the Laplacian, Green's functions, Poisson brackets, and similar topics. It seemed sometimes that Volterra did not want to reveal the physics underlying the equations and the analogies between different theories (Segrè 1993).

Besides "institutional" lectures, Majorana also attended "private" lessons held by Fermi in his office for his students and collaborators:

> Fermi gave a course in which he explained exactly what is contained in "Introduzione alla fisica atomica". [...] He wrote [this book] in 1927. He taught exactly that. Then, I would say three times a week, in the afternoon at five o'clock or something like this, he would give us a private lecture. To Amaldi, to me, to Rasetti, and Majorana, and occasionally Corbino would come. Not everybody always came, but I did and so did Amaldi; Majorana, yes, but very often he would say, "Well, it's beneath my dignity. Why should I learn these things? You are doing it in a childish way; it should be done this way." [...] Fermi didn't react to that. Except a year or two later he decided that we were not worthy enough to be present at the interviews [when] Majorana [was there], and then they would closet themselves together, you see, because they went very fast in their discussions of very difficult theory.[25]

---

[25]T.S. Kuhn's interview with Segrè, *loc. cit.*

During the fourth year of the physics course, Majorana was also preparing his degree thesis, and of course Fermi was his supervisor. It is interesting to note how, as a matter of fact, all the exams at the Institute of Physics were given less than fifteen days before his graduation. His thesis dealt with the theory of radioactive nuclei, which might be considered unique: at the time, Fermi's group was working mainly on atomic and molecular physics, not yet nuclear physics, which would only become their main topic of research from 1933 onward. But, as Segrè recalls, this was not at all unusual:

> Fermi did not like to assign subjects for doctoral dissertations, or in general to suggest subjects of investigation. He expected the students to find one by themselves or to obtain one from some colleague who was more advanced in his studies. The reason for this, as he later told me, was that he did not easily find subjects simple enough for beginners: he generally thought of problems that interested him personally and were too difficult for students (Segrè 1970).

And so the first work on nuclear physics in Fermi's Rome group was carried out by Majorana himself, as a student. Once again, what stands out is the insightfulness of this work, together with his mastery of the scientific literature on the topic, as is also clear from his *Volumetti* and *Quaderni*, kept in Pisa.

On July 6, 1929 Ettore defended his master's thesis before a committee chaired, among others, by Corbino, Fermi, Volterra, Levi-Civita, Lo Surdo, Armellini, and Trabacchi. The title was *The quantum theory of radioactive nuclei*; the accompanying oral "papers" were: *On a photoelectric effect in "audions"*; *On the equilibrium configuration of a rotating fluid*, and *On statistical correlations*. The reasonably expected result was *110/110 magna cum laude*.

What many would consider a good starting point, Majorana, who was eager to deal with "pure science",[26] simply considered a transition phase, as we will see later on.

---

[26]Résumé sent by Majorana to C.N.R., dated May 1932; quoted in (Recami 1987).

Chapter 2
A Certain Interest in Pure Science

Theoretical Research

As soon as Ettore Majorana moved to Fermi's group, he distinguished himself not only as an expert calculator but also by his extraordinary abilities as a researcher. In the Rome group some seminars were organized periodically on specific topics, these being held in rotation by the different members of the group. When it came to Majorana's turn, although everybody was very attentive, only Fermi was able to understand what was presented, and sometimes even showed his "irritation" towards Ettore, accusing him of "not saying everything", and would invite him to complete his presentation.[1] Actually, according to accounts by some of those present, the only one in the group to hold his own with Majorana was Fermi. From the start, the relationship between them was as peer to peer, also because of the small age gap. An episode recalled by the chemist Oscar d'Agostino, who also joined the *Via Panisperna* group (although later), and who made notable contributions to some of their research, is illuminating. One day some students, who were coming back to the Institute in the afternoon, after the lunch break, found Fermi and Majorana in a lecture hall in front of blackboards full of calculations, shouting at each other in a lively argument. "They were calling each other fools. The argument had started at midday and they had been there several hours in a passionate discussion, forgetting of course to eat lunch".[2]

Well before changing to physics, while he was still attending the Engineering School, Majorana often used to adopt a "personal" view about what was discussed in those university lectures that particularly struck him. Examples of this habit can be found in certain manuscripts, in particular, the already mentioned "Volumetti" (five notebooks totalling about 500 written pages), now kept in Pisa. It is clear that some of the topics presented in the lessons served as a starting point for personal

---

[1]This story was provided by the engineer Gabriele Paparo, one of Amaldi's former collaborators; he used to tell his collaborators anecdotes about his friend Ettore during breaks at work.
[2]See the article (Ferrieri and Magnano 1972).

© Springer International Publishing AG 2017
S. Esposito, *Ettore Majorana*, Springer Biographies,
DOI 10.1007/978-3-319-54319-2_2

research later on. Anyway, Majorana's interests were not only of the academic kind, and a nice example is given by the anecdote in the previous chapter about the meeting between Majorana and Fermi. Indeed, in the Pisa manuscripts we can find the pages where Ettore wrote down the solution to the Thomas-Fermi equation. This served as a basis for the study of many atomic problems, one of the main topics of physics research from that period to which Fermi's group also contributed.

And Fermi himself, noticing the results achieved by Majorana in their discussions, invited him to present them (Esposito 2005a) in a talk at the XXII General Assembly of the Italian Physical Society (which took place at the Institute of Physics in Rome in 29–30 December 1928), long before he graduated in physics (and on a different topic to the one he discussed at his graduation). Even then, however, the physicist from Catania had a tendency not to publish the results of his research, when these could be considered premature, or at least not until they had reached an adequate form, according to his overcritical judgement, of course. As a matter of fact, Majorana actually gave a report at the said conference on December 29, between two of Fermi's speeches, but he did not regularly publish his work: *"the research carried out so far is still insufficient to appreciate the full value of these results"*, he said at the end of his talk.[3]

The question of the Thomas-Fermi atomic model must have particularly fascinated Ettore, if it is true that he actually applied "his" solution to other particular problems (only in a few cases did Majorana later return to specific topics in his *Volumetti*, as indeed he did on this occasion), sometimes with the clear purpose of re-deriving results already obtained by Fermi. An application of the statistical method by Thomas-Fermi to atoms was also the object of the study reported in Majorana's first paper, published with Giovannino Gentile a year before graduation. This is the only proof of the collaboration with his friend Gentile, who frequented the Fermi group in Rome at that time (for the other papers, Majorana was the only author). However, a close friendship developed between the two of them, as attested in their exchange of letters during the period when they were not working in the same establishment, and Majorana never missed an opportunity to help his friend with particular questions of physics and mathematics.

As we have seen, Majorana's way of promptly answering requests from friends and colleagues who were experiencing difficulties in solving certain scientific problems dates back to his university years in engineering. However, it remained a feature during the following years in research as well. Segrè provides a typical example of this: in 1931–2 he was on a fellowship in Hamburg (Segrè 1993), where he was working with Otto Robert Frisch in Otto Stern's group, carrying out experiments on atoms in magnetic fields. To interpret the results of the experiment they needed to solve some theoretical problems and, perhaps during a holiday break in Rome,[4] Segrè involved his friend Majorana (Amaldi 1968). The latter duly came

---

[3]See what is quoted in *Il Nuovo Cimento*, 6, XIV–XVI.

[4]T.S. Kuhn's interview with E. Segrè, *loc. cit.*

up with the appropriate theory for the phenomenon being studied experimentally by Frisch and Segrè, and it became his sixth published article.

Another revealing example of Majorana's approach is the following. Toward the end of 1931, after a few months in Leipzig (where he had been working with Peter Debye), Edoardo Amaldi was conducting some experiments on the spectrum of the ammonia molecule together with George Placzek, who was then a visiting scientist at the Institute of Physics in Rome. In those days, after spending a few hours of the afternoon in the library, Majorana would join Amaldi in the laboratory, and when work was over, they would go back home together (the two friends lived quite close to each other). During these walks, besides making friends with the Czech visitor, Ettore learned about the research they were doing in the lab. And these investigations certainly caught his interest, so much so that he solved the problem of determining the ammonia oscillation frequencies on his own, relating them to the geometrical structure of the molecule, which has a tetrahedral shape. This study was never published by Majorana (Di Grezia 2006), but was stored in his personal notebooks, and it was probably never revealed to his friends, because they make no mention of it.

We also have the observations of a newcomer to the group, Gian Carlo Wick, who had just joined Fermi in the autumn of 1932, and who remembers this of the few contacts he had with Majorana:

> [They] were enough to understand the extraordinary versatility and quickness of the workings of his mind. I also noticed his hypercritical but slightly ironical attitude towards, not only the others, but also his own work. Still he was too kind and fair to exercise his irony on someone who was inexperienced, and on the contrary he showed his interest in what I was doing with encouraging words – it concerned the magnetic moment of a hydrogen molecule in its various states of rotation. He also gave me some useful advice about some numerical calculations required by the problem, suggesting the use of a function different from the Heitler-London function I had been using. After hearing what value I had got from the first calculations, he observed: "Yes, this may be an upper limit; the other function will give a lower limit, which will be remarkably smaller". I was quite surprised: how could he know that? That was Majorana's famous intuition! But when he explained his reasons, I became persuaded that they were totally rational. Needless to say, the conjecture proved to be perfectly right (Wick 1981).

He was equally open, not only towards his own friends, but also with the young international guests who came and worked for a while with Fermi. For example, one of these guests, Rudolph E. Peierls, remembers:

> I surely received a lot of useful ideas and explanations from Fermi and the others, as Wick and Majorana [...]. Majorana was quite strange and retiring; he was a Sicilian [...]. He made his reputation on two important things. One the exchange nature of the nuclear forces, where he basically corrected an oversight in Heisenberg's ideas. And then the other was the neutrino theory.[5]

There were also times when Majorana's precious contribution was put to good use by Fermi himself in his research, as happened in 1932 when he and Segrè were elaborating the theory of the so-called hyperfine structures of atomic spectra. In the

---

[5]T.S. Kuhn's interview with R.E. Peierls, 18 June 1963.

work published by Fermi and Segrè in 1933 (Fermi and Segrè 1933), they explicitly thank "*Dr. Majorana for several discussions about calculations*" needed for their theory; examples of Majorana's calculations on this topic can be found in the pages of his *Quaderni*, kept in Pisa.

For this reason another episode reported by Amaldi[6] is particularly enlightening. On Sundays during the summer, Fermi and the other members of his group used to go to Fregene (a popular seaside resort) and Majorana would join them. However, he was always late, as he went there by bus from Rome, unlike the others. Once Majorana told his friends that the night before he had been working on some problem of quantum electrodynamics (in his personal notebooks there are, in fact, many pages devoted to this topic), but that he had "not quite figured it out". However, he informed them that he had overcome the theoretical difficulties right that morning, on the bus to Fregene, and had written down the related calculations on a cigarette packet. He then proceeded to show the theory to his friends (who may well not have understood it at all, as Amaldi recalled), and wrote the formulae with his finger in the sand. Anyway Majorana never went back to it again, and the work was finally carried out by Fermi.

Ettore's theoretical contribution to Fermi's group in Rome was therefore substantial, but most of his research and the most relevant part was what he carried out *motu proprio*, and this can be found in his personal study notes. Particularly important was Majorana's discovery of Hermann Weyl's book on the mathematical theory of groups applied to quantum mechanics (Weyl 1928). This was in 1928–9 and it deeply affected all the later work of the physicist from Catania. To his friend Gentile, who was in Germany, he wrote on 22 December 1929:

> As for me, I am not doing anything sensible In fact, I am studying the theory of groups with the firm intention of learning it, which makes me a bit like that hero from Dostoyevsky who, one fine day, started saving loose change, in the hope that he would soon be as rich as Rothschild.[7]

The concise, clear, and general view given by the theory of groups was, as a matter of fact, well appreciated by Majorana (and by very few others in the world, until its eventual rediscovery during the 1950s–1960s), and would form the basis for all his most important work.

The "Neutral Proton" and the Turning Point in 1932

James Chadwick's discovery of the neutron, the companion particle of the proton in the atomic nucleus, was made at the beginning of 1932. It proved to be a turning point for nuclear physics around the world, and hence also for Fermi's group in Rome. The actual experimental evidence came in the wake of several intuitions which later turned out to be false: these tried to explain certain strange observations

---

[6]This anecdote too was recalled by the engineer Gabriele Paparo, already mentioned.

[7]Letter MG/R1 of 22 December 1929 in (Recami 1987).

of nuclear reactions without postulating the existence of an electrically neutral subnuclear particle (De Gregorio 2007), namely, what we now know as the neutron.

> When Joliot's paper arrived in Rome in January (1932) with the Comptes Rendus and the paper of Chadwick had not yet appeared, we all were very much interested in it. But Majorana said: 'How stupid of Joliot! They have not understood that this is the neutron". […] "This is obvious. These gamma rays make no sense in a nucleus. They should be neutral particles". Well, we used the word 'neutron' because, you remember maybe, the idea of a neutron was suggested by Rutherford a few years before in order to explain the anomalous scattering of alpha particles. Gentile, another young man of our group, a great friend of Majorana's, had done some calculations trying to introduce in quantum mechanics the effect of neutral particles moving around the nucleus. He had taken the idea of Rutherford and he had done some calculations. So the idea of a neutral particle, suggested by Rutherford, was discussed in the Institute. […] The comments of Majorana were really quite interesting. He would say, "Well, how stupid[8] they are!" This was not because they were stupid, because they were very intelligent people – it was just his way of expressing himself. "They don't understand! This should be a neutral particle. There's no sense to think of gamma rays of 50 MeV. There is no sense to talk of gamma rays of 50 MeV. This should be neutral". […] No, nobody took Majorana's suggestion seriously. […] Majorana said it then, but everybody said, "Ha, ha". […] When Chadwick's paper came out, we were all convinced. And we said, "Look how quick Ettore is – he understood before Chadwick".[9]

We have proof of Majorana's intuition, though indirectly, from R.E. Peierls, an international guest in Fermi's group:

> This of course was just the time when the artificial radioactivity had been discovered, and when there were the experiments beginning to come out which led to the discovery of the neutron. Fermi always had a slightly peculiar attitude to that. I think he felt that the Paris group, the Joliots, should really have seen the existence of the neutron from their experiments which were later pointed out by Chadwick. I had the impression that he knew what the experiments meant, but hadn't got round to publishing it, or felt he must leave it to the experimenters. I don't know. This is only a hunch.[10]

Majorana's happy intuition, which remained unknown outside the Rome circle, was not a mere improvisation! As usual, it was based on a deep knowledge of nuclear physics. This proficiency dated back to his time at university (his graduation thesis dealt with the mechanics of radioactive nuclei) and it continued in the following years, as we can see from the pages of his *Volumetti* (Esposito et al. 2003) and *Quaderni* (Esposito et al. 2008).

As Segrè recalls, the group at the *Via Panisperna* Institute had in fact long been discussing the possibility of changing tack and focusing on nuclear rather than atomic physics (Segrè 1993). While Fermi was quite ready to begin seriously

---

[8]Amaldi himself tells us that: [Majorana] felt everybody was stupid, himself included. The only people who were not completely stupid according to him were Dirac and Weyl. He greatly respected Weyl. Heisenberg was stupid. Everyone else was stupid. Even himself. He would not say that the others were stupid and he was clever. No, he was convinced that he was a fool. […] Pauli was rather stupid. But Majorana was stupid as well, he was convinced of that. This gave him a feeling of emptiness.

[9]T.S. Kuhn's interview with E. Amaldi, 8 April 1963.

[10]T.S. Kuhn's interview with R.E. Peierls, *loc. cit.*

studying the nucleus, his young colleagues were doubtful about abandoning the research they had by then become quite expert about, especially since they had only recently mastered its most advanced experimental techniques. However the general orientation of the group gradually changed, even though they did not completely abandon the study of atomic phenomena;[11] as a "bridge" between the two different fields of physics, they started to investigate the problem of hyperfine structures, as mentioned above.

Chadwick's discovery, announced in February 1932, shook the international community, and many physicists devoted themselves to develop theories on the structure of nuclei which would also take the neutron into account as a constituent. The aim was to interpret the mounting experimental evidence in a satisfactory (and correct) way. Amaldi remembers that "before Easter that same year, Ettore Majorana had already tried to develop a theory of light nuclei assuming that they were made up only of protons and neutrons (or "neutral protons" as he used to say then) and that the former interacted with the latter through exchange forces (Amaldi 1968). Such forces are of a purely quantum origin and are responsible for the existence of certain kinds of molecules, as had been discovered by Heitler and London at the end of the 1920s; Majorana already had a good knowledge of them, as attested by some of his papers published in the previous years and in his personal study notes.

> Immediately after Chadwick's discovery [...] Majorana wrote down the expression for all possible forms of the exchange forces. All of them. Then he started to make a shell model; this is not known. He started to calculate the [energies of the] shells, but when he reached carbon, he was unable to go any further. He had started to do the calculations for helium [...] but when he reached carbon, or immediately after carbon – and these are very light elements – it suddenly became difficult. There is nothing astonishing about that. But he said that was stupid. It was obviously complete nonsense.[12]

Majorana "had told his friends at the Institute about this rough theory" and, as recalled by Eugene Feenberg, who was in Rome with the others at the time:

> [...] he actually gave a seminar on his forces, nuclear forces. I remember Uhlenbeck and Inglis were there at that time and they thought it was a very big thing.[13]

"Fermi, who immediately understood their importance, suggested that he should publish the results as soon as possible, even if they were incomplete. But Ettore did not want to hear about it, as he considered his work incomplete" (Amaldi 1968). Needless to say, what Majorana's overcritical eye viewed as "incomplete" was not

---

[11]Another example of the incipient shift in interest in Italy is provided by the Institute of Physics in Naples, directed by Antonio Carrelli. Perhaps drawn by Fermi's euphoria, Carrelli originally took an interest in the physics of the nucleus, but the main topic of research conducted in his institute (and later, the only one) remained atomic and molecular spectroscopy. This decision, possibly due to the limited means available, did not take the same happy turn as the Rome group.

[12]T.S. Kuhn's interview with E. Amaldi, *loc. cit.*

[13]C. Weiner's interview with E. Feenberg, 13 April 1973, kept at the Niels Bohr Library of the American Institute of Physics.

at all considered as such, even by a first-class physicist like Fermi, and this is once again confirmed by the notes kept in Pisa (the *Volumetti* and particularly the *Quaderni*). As so rightly pointed out (De Gregorio 2007), Fermi and his group, with the clear and decisive support of Majorana, contributed substantially within the international scientific community to the emergence of the idea (which later proved to be true) of the neutron as an independent and fundamental particle, a "neutral proton" and not a combination of a proton and an electron.

Despite Fermi's encouragements, Majorana never published anything about his theory on the structure of nuclei in terms of just protons and neutrons. When Fermi was invited to attend an important conference in Paris on 7 July 1932, he chose to take stock of the situation in the physics of the atomic nucleus, and again decided to put pressure on Majorana, sure as he was of the relevance of the theory Ettore had developed. So, according to Amaldi, "he asked Majorana's permission to sketch his ideas on nuclear forces. Majorana forbade Fermi to speak about it, saying that, if he did want to, he might do so, but in that case he should say they were the ideas of a famous professor of electrical engineering,[14] who happened to be at the Paris Conference and whom Majorana considered a living example of how not to do scientific research. The condition imposed by Majorana had the clear purpose of preventing Fermi from quoting his "incomplete" theory. His allusion to the "famous professor of electrical engineering" relates to the personal vicissitudes of a cousin of his, with whom Ettore was in close contact, and who was then struggling to pass his own electrical engineering exam at the *Scuola degli Ingegneri* in Rome (his cousin eventually gave up his studies).[15]

Some weeks after Fermi came back from the conference, Werner Heisenberg's first paper on nuclear theory appeared: it was based on the idea of an "exchange force" among nuclear constituents, already successfully introduced in the theory of the molecular bond, and made it possible to solve many theoretical difficulties in the understanding of nuclear structure. The international scientific community soon understood that, even though it was incomplete and there were some undeniable imperfections in Heisenberg's theory, they were at last on the right track. "In the Institute of Physics at the University of Rome, everybody was extremely interested and full of admiration for Heisenberg's results, but at the same time they were disappointed that (Majorana) had neither wanted to publish, nor allowed Fermi to speak of his ideas at an international conference. Fermi did his best once again to convince Majorana to publish something, but every effort of his and of ours, his friends and colleagues, was useless. Ettore would answer that Heisenberg had already said everything that could possibly be said and that, furthermore, he had probably said too much".

---

[14]The "professor of electrotechnics", as Amaldi recalls, was Giovanni Giorgi, known for his system of physical units.

[15]This piece of news was referred to us by A. De Gregorio.

This last caveat clearly refers to the flaws in the model proposed by Heisenberg and which prevented him from going further, flaws that were not present in Majorana's theory, as we will see in the next chapter. As a matter of fact, the idea of an "exchange force" which held together protons and neutrons inside the nuclei, as conceived by Heisenberg in close analogy with what happens between two hydrogen atoms H held together in the H_2^+ molecular ion by the exchange of an electron, actually contained within it the idea that the neutron was somehow made up of a proton and an electron. In contrast, Majorana's idea of a "neutral proton" led to the consequence that the nuclear bond was not due to the actual exchange of an electron among protons and neutrons in the nucleus, but rather to the exchange (according to quantum mechanics) in the mutual positions of the protons and neutrons. The consequences of Majorana's theory, unlike those of Heisenberg's, were well supported by experimental facts, and Majorana was clearly aware of this in his hint that Heisenberg might have said "too much".

A Trip to Leipzig

Shortly after what happened in July 1932, Fermi insisted once again: on the one hand he urged Majorana to obtain a lecturing post in theoretical physics (to teach in Italian universities), while on the other he encouraged him to go abroad, perhaps to visit Heisenberg himself. This time, in contrast to his other attempts, Fermi succeeded on both fronts: Ettore competed for, and as expected, obtained a lecturing post—*libera docenza*—in November 1932 (as we will see later on), and in the meanwhile he was persuaded to make a journey abroad, perhaps following the example of other members of the Rome group (Ettore's friends such as Giovannino Gentile, Edoardo Amaldi, Emilio Segrè, and others). As soon as he got his approval, Fermi managed to obtain a grant for this trip from the *Consiglio Nazionale delle Ricerche* (C.N.R.).

The "programme" for his trip abroad is clearly stated by Majorana himself in his letter dated 9 January 1932 to the vice president U. Bordoni of the National Committee for Astronomy, Physics and Applied Mathematics of the C.N.R. (Recami 1987):

> It is my intention to leave around the 15th of this month for Leipzig and stay there the whole month of June to carry out, under the guidance of professor W. Heisenberg, theoretical research focusing mainly on the structure of nuclei and the relativistic formulation of the new quantum theory. During the holiday period between the winter and summer terms I will be returning to Italy for about fifteen days. I will spend the rest of the time attending conferences and scientific meetings which traditionally take place during that period in Germany and Denmark.

Once he got the C.N.R. grant, Majorana eventually left for Leipzig, where Heisenberg had brought together a group of young and very talented physicists, making the Institute of Theoretical Physics one of the major world centres for the study of modern physics. The productive environment that welcomed Ettore in Germany is well described by his colleague Giovannino Gentile, who had been there a couple of years before, and who probably described it to his friend, as Ettore's first two letters[16] to Giovannino seem to attest:

From Leipzig, where I am staying for my studies, I am going back to Berlin after almost a year. This time I am leaving with my colleagues. Heisenberg is with us too and we are here to meet our colleagues in Berlin. These meetings are interesting. We talk, we discuss things, we get to know each other; that is the way we work together. [...] In Leipzig, in our Institute of Theoretical Physics, there is a bunch of people, from different countries, who work around a man who has brought brand new ideas and methods to science. This man is Werner Heisenberg. Very young, about twenty-nine or thirty, fair haired, he almost looks like an Englishman: but in fact he is German, a rather typical German, with a typical mentality and refined culture. [...]

But it is mainly in the "colloquia" – weekly meetings held in the various physics institutes – that one feels the greatness of Berlin scientific life. After all this is an institution common to all the faculties of every German university. [...]

What I am saying is that it is hard to put across how productive such conversations, and such collaboration, can be. There is no such thing as an isolated scientist here; everywhere he will hear voices echoing his own, and we are easily motivated to scientific research and critique, which is the first condition for carrying out any scientific work. Because any science not yet crystallised in a research paper will pulsate and shake, and a new thought can arise spontaneously in such luxuriant surroundings.

This is one of the reasons why here in Germany one is soon convinced that the conditions for scientific research are extremely favourable, precisely because there is such a wealth of men and equipment.[17]

Ettore arrived in Leipzig on the night of 19 January 1933,[18] and two days later he was already talking about[19] "the work to be done" with Heisenberg. To his mother he wrote (on January 22):

At the Institute of Physics I have been warmly welcomed. I have had a long conversation with Heisenberg, who is an extraordinarily polite and friendly person. I get on very well with everybody, especially with the American Inglis whom I had met in Rome and who now often keeps me company and shows me around.[20]

During his stay abroad, Majorana maintained "personal relations with various eminent people";[21] among others, he met D.R. Inglis, E. Feenberg, F. Bloch, P. Ehrenfest, F. Hund, P. Debye, B.L. van der Waerden, G. Wataghin, G. Placzeck, N. Bohr, C. Møller, V. Weisskopf, H. Kopfermann, W. Pauli, H.A. Bethe, L.

---

[16]Letters MG/R1 of 22 December 1929 and MG/R2 of 15 May 1930 in Recami (1987).

[17]G. Gentile jr.'s letter to his family of 9 February 1931, in Gentile (1942).

[18]Letter MF/L1 of 20 January 1933 in Recami (1987).

[19]Letter MB/L1 of 21 January 1933 in Recami (1987).

[20]Letter MF/L2 of 22 January 1933 in Recami (1987).

[21]Letter MF/L4 of 14 February 1933 in Recami (1987).

Rosenfeld, and G. Beck. He soon felt the difference in the working conditions as compared to the Institute of Physics in Rome, these being clearly better suited to his research in theoretical physics. As a matter of fact, where Fermi had failed with Majorana, Heisenberg now easily succeeded just a few weeks after his arrival: "I am writing some papers in German. The first one is ready".[22] And, even more explicitly: "I have written a paper on the structure of nuclei which Heisenberg much appreciates, though it contained some corrections to a theory of his".[23] We already dealt with these remarkable corrections in the previous paragraph, but it is quite surprising that Heisenberg himself (who would receive the Nobel prize right at the end of 1933) stressed the importance of Majorana's theory on many occasions, as Ettore had already noted in a letter to his family:

> In our last "colloqium", a weekly meeting among more or less a hundred physicists, mathematicians, chemists, etc., Heisenberg spoke about the theory of nuclei and sponsored one of the works I have written here.[24]

On this occasion, Heisenberg also invited Majorana to present his theory to those present, but he refused:

> I remember a conversation when Heisenberg was talking about nuclear forces. Majorana was among the listeners and Heisenberg asked him to stand up and explain his version of exchange forces.

> But I also remember that Majorana was too shy or too aware of his lack of knowledge of the German language, whence I am quite sure that he declined the invitation and said nothing.[25]

Even after this episode, as for example at the VII International Solvay Conference of Physics held in Brussels in the autumn of 1933, during his lecture on the theory of nuclear forces, Heisenberg "*almost always* uses sentences like "according to Majorana", "following the example of Majorana", "as Majorana has stressed", "we will adopt, as did Majorana", mentioning his personal contribution only on rare occasions and in a very understated way. This reference to the Sicilian physicist is so persistent that he seems to want to express a debt of gratitude towards the colleague who had made some corrections to his theory" (De Gregorio 2007).

The general impression Ettore gave of himself in the Leipzig group is reported by Heisenberg himself during an interview of 1963 archived in the Oral History of Quantum Physics:

> There was a very good physicist in my laboratory in Leipzig with the name Majorana. You know the so-called Majorana representation of the Dirac particle. He came as a young Italian physicist to Leipzig. He was a very brilliant man and at the same time a very nervous type of a man. He did excellent work. He was always extremely pessimistic about physics. I tried always to induce him to write papers and so he did finally write a very good paper.[26]

---

[22]Letter MF/L4, *loc. cit.*

[23]Letter MF/L5 of 18 February 1933 in Recami (1987).

[24]Letter MF/L6 of 22 February 1933 in Recami (1987).

[25]C. Weiner's interview with E. Feenberg, *loc. cit.*

[26]T.S. Kuhn's interview with W. Heisenberg, 28 February 1963, Session no. 10.

Heisenberg was clearly struck by Ettore's pessimistic character:

I would say that he was perhaps not pessimistic about physics especially, but rather about life in general. He was that kind of difficult fellow. Well, sometimes I thought perhaps he had had very difficult experiences in his life with other people, perhaps with girls or something like that. I don't know. Anyway, I couldn't make out why he, being such a young man, and such a brilliant young man, could always be so pessimistic. He was a very attractive fellow, so I liked him in our Leipzig group. I tried to see him frequently, and we had him with us for our ping pong games. Then I would sit down with him and ask him, not only about physics but more personal things, and so on. So I tried to keep in touch with him. He was a very attractive fellow but very nervous, so he would get into a state of some excitement if you talked to him. So he was a bit difficult.[27]

And again:

People tried to talk to him and he was always very kind and very polite and very shy. It was very difficult to get something out of him. But still, one could see at once that he was a very good physicist. When he made a remark, it was always to the point.[28]

The relationship between Majorana and Heisenberg struck many of those who witnessed their conversations, as one of these remembers, A. Recknagel:

I was very much impressed to see how Majorana would discuss with Professor Heisenberg as an equal. At the time I was still a student, and what a professor said was the truth for me, all the more so if that professor had been awarded a Nobel prize. Only gradually have I understood that you can criticize a professor the same way you criticize a student. But at the time if a young man like Majorana was able to discuss freely or even criticize a professor, that seemed astonishing to me. And that is the reason why I still remember Majorana today.[29]

The theory of nuclei, however, was not the only topic Ettore was developing during his stay abroad. For instance, Victor Weisskopf tells us that he had "a discussion with Majorana about the latest developments in quantum electrodynamics",[30] a topic which was to become one of Ettore's favourites, as we will see later on. However, the most interesting news concerned the pioneering paper on the "relativistic theory of particles with arbitrary intrinsic momentum", published before he left for Leipzig and which we have already discussed.

In this paper there is an important mathematical discovery, as I have checked with professor van der Waerden, a Dutchman who teaches here, one of the greatest authorities on the theory of groups.[31]

---

[27]T.S. Kuhn's interview to W. Heisenberg, *loc. cit.*

[28]T.S. Kuhn's interview to W. Heisenberg, 5 July 1963, Session no. 11.

[29]Interview to A. Recknagel in F. and D. Dubini, *La scomparsa di Ettore Majorana*, television programme aired in 1987 by Swiss TV.

[30]Testimony T4 in Recami (1987).

[31]Letter MF/L5, *loc. cit.*

The next episode, told by Heisenberg, is also revealing and it involves B.L. van der Waerden:

> van der Waerden's role in Leipzig was very important because he had a tremendous ability to understand quickly what people were talking about [concerning group theory] and then he knew all these things so well, so with just a few sentences of explanation, he could immediately clarify a complicated situation arising at one of our seminars. [...] I feel that I have really learned a large part of my mathematical training from van der Waerden, just by discussing with him. [...] Well, we spoke about the Weyl spinor business. The Dirac spinor was the thing everybody was talking about, but then there was the Weyl spinor business, which van der Waerden knew. The others did not know about it, but then there was Majorana. He was in Leipzig and Majorana found his Majorana particle, which has no charge but still has spin 1/2, and that, of course, had to be represented by Weyl's spinor.[32]

Clearly encouraged to go further in this direction (unlike the theory of nuclear structure, Fermi and the other members of the Rome group had not understood the importance of this work which was, without doubt, among the author's most theoretical), Majorana wrote to professor Bordoni (C.N.R.) to inform him that "the manuscript for a new relativistic theory of elementary particles is ready".[33] This paper is now lost (perhaps, according to the above quote from Heisenberg, it contained the theory Majorana published some years later on the so-called "Majorana neutrino"), and it was certainly not published in the German journal as had been Ettore's intention, probably under pressure from C.N.R. bureaucrats who wanted papers to be published first in Italian journals (Recami 1987).

Anyway, his personal studies continued in this field too, and the subject matter became one of those he spent most time investigating, even after his return to Italy, as is clear from the *Quaderni* kept in Pisa, where many pages are devoted to it. Here his interest in purely theoretical questions (but those with a noted experimental relevance), linked both to quantum electrodynamics and in particular to Dirac's theory of the electron, is quite evident. Although it had already developed some time earlier during his personal studies in the library of the Institute of Physics in Rome (where Ettore would spend his time reading papers by Heisenberg, Dirac, and others), this "passion" really flourished in the favourable theoretical environment in Leipzig. As a matter of fact he often hinted at that in his letters to friends and colleagues (such as Segrè and Gentile):

> They take Dirac's theory of positive electrons quite seriously. Heisenberg studies the properties of relativistic invariance and the possibility of having other applications, besides Dirac's calculation of the lifetime of positive electrons. Of particular importance is the calculation, already attempted by Beck, of the probability that a high-energy light quantum produce a pair of oppositely charged electrons when scattering from a heavy nucleus.[34]

The winter term in Leipzig finished at the end of February, after which there were almost two months of holiday; Majorana took advantage of this to go to

---

[32]T.S. Kuhn's interview with W. Heisenberg, 5 July 1936, *loc. cit.*

[33]Letter MB/L2 of 3 March 1933 in Recami (1987).

[34]Letter MG/L1 of 7 June 1933 in Recami (1987).

Copenhagen "until April 15 in the company of professor N. Bohr",[35] as in the "schedule" he had presented some months earlier. However, to begin with, perhaps inspired by Heisenberg, on February 7 he wrote to his family:

> I am going to stay in Leipzig until the end of February. In March and April, there are holidays here. I will probably take advantage of this by going to Zurich work with Pauli, one of the greatest living scientists.[36]

As Majorana himself would state when he wrote to his friend Gentile from Copenhagen,[37] "there does not seem to be much choice for theoretical physicists outside Leipzig, Zurich, Copenhagen, and Rome". These were of course the places where Heisenberg, Pauli, Bohr, and Fermi were working. It is curious to note that, though keen to move to Copenhagen for a few weeks, Ettore found it hard to leave the stimulating German environment that he had so recently joined:

> I am really sorry I have to leave Leipzig where I was warmly welcomed, and I shall be glad to return in two months.[38]

And again:

> I may stay in Leipzig for two or three days more because I must talk with Heisenberg. His company is unique and I wish to take advantage as long as he is here.[39]

Ettore arrived in Copenhagen on March 4 and fitted in "perfectly from the first moment".[40] Even if his meeting with Niels Bohr, one of the founding fathers of quantum theory, and the other guests at Bohr's *Institut for Teoretisk Fysik*, was pleasant enough, it is easy to tell from Majorana's letters that the new experience in Denmark was not so fruitful as the one in Leipzig, to which he enthusiastically returned on May 5, after a short trip to Italy for the Easter holidays.

He soon got back to his old study schedule, but with less intensity, owing to a persistent gastritis whose symptoms where hard to bear: "my activity in the last month has been somewhat reduced due to my poor state of health".[41] This illness would have certain consequences on Ettore's future activity as well, but in the final reckoning his stay abroad could only be considered highly positive.

---

[35]Letter MB/L2, *loc. cit.*

[36]Letter MF/L3 of 7 February 1933 in Recami (1987).

[37]Letter MG/C1 of 12 March 1933 in Recami (1987).

[38]Letter MF/L6, *loc. cit.*

[39]Letter MF/L7 of 28 February 1933 in Recami (1987).

[40]Letter MG/C1, *loc. cit.*

[41]Letter MG/L1, *loc. cit.*

Back in Rome

Once back in Italy in the summer of 1933, Majorana entered several "gloomy years" of isolation. Amaldi, for instance, says (Amaldi 1968):

> He started attending the Institute in *Via Panisperna* only occasionally and, as the months went by, he stopped coming at all; he would spend his days at home, immersed in his studies for hours on end.

And again:

> More than one attempt made by Giovanni Gentile Jr, Emilio Segrè, and myself to bring him back to a normal life was useless. [...] None of us was able to discover whether he was still keen on theoretical physics; I think so, but have no evidence.

We shall not dwell further on the possible reasons why he acted this way: it might have been his poor state of health since the last part of his trip abroad or, later on, his father's death in 1934. We just don't have enough information (but the reader is free to consult the available bibliography).

It is far more interesting to linger on his possible scientific and/or even educational research. Even to a witness like Amaldi, it sounded strange that Majorana would so suddenly abandon theoretical physics to devote himself to topics such as "politics, the navies of different countries and their balance of power, and ship-building characteristics" (Amaldi 1968), or indeed to philosophy—although that was something he had always cultivated—and even medicine. Indeed, he had always had a deep interest in theoretical physics over the previous years (even before his graduation), a fact proved not so much by his publications as by his many personal notes. Nevertheless, all of these and other "strange" topics were among the subjects Ettore took up in those years, as we can see by the presence of the Italian Nautical Almanac of July 1937 in his otherwise scanty personal library (only 29 volumes, later looked after by his sister Maria). And the "attempts" Amaldi refers to are indirectly documented by the presence of at least two or three books in his library: James Jeans's *I nuovi orizzonti della scienza* [*The New Background of Science*] (1934), containing a handwritten inscription by his friend Giovannino Gentile, who was the Italian editor of the book; Franco Rasetti's *Il nucleo atomico* [*The atomic nucleus*] published in 1936; and *Fisica nucleare* [*Nuclear Physics*] written by Gentile and published in 1937, also autographed by the author. His friend Giovannino's two volumes were probably not given to Ettore in person, as there are two thankyou letters, the first dated 27 July 1934, and the second 20 June 1937. Clearly, such letters seem to contradict the idea that Majorana had abandoned physics, as they testify to a non-superficial reading of the two texts. Even the circumstance that these letters were written at Monteporzio Catone, where the Majoranas used to spend the occasional holiday breaks, tends to contradict the idea of Ettore's living in "isolation": they were both written in summertime, while other letters he wrote testify to his presence in Rome at various times of the year. It was perhaps rather the serenity of the country house that favoured his reading the cited texts.

In his first letter of 27 July 1934 Majorana wrote:

I thank you very much for sending me Jeans's book in your nice edition (and translation?).
It arrived at a perfect time to distract me from my country idleness. I admire the thoughtful
preface, well suited to Italian readers, with appropriate references to the prevailing current
of thought here. I think that the greatest quality of this book is that it anticipates the
psychological reactions the recent advances in physics will necessarily produce once it is
generally understood that science has stopped being a justification for vulgar materialism.
I do think your translation can seriously contribute to reviving an interest in scientific
problems here in Italy. I will be in Rome in a few days and I hope we will have the chance
to meet.[42]

The hint about the "materialistic" use of science is clear evidence of Majorana's
interest for philosophical questions, but his main concern nevertheless always
seems to be an "interest for scientific problems".

Likewise, but in an even more direct tone, in the letter dated 20 June 1937:

I do thank you for your fine book *Fisica nucleare*. It is a truly perfect educational work, and
for the amount and quantity of information, it is indeed an agreeable read, extremely
interesting for anyone with a minimum of technical competence. I hope your publisher can
"launch" it properly, because there has not been anything like that in Italy for ages, and nor
will there be any time soon. It really should be in everybody's hands.[43]

Finally, it is worth noting that, at least according to the documents we currently
have in our possession (above all Rasetti's book, as Gentile's second book might
simply have been a gift from a friend), Majorana had not apparently spent much
time doing theoretical physics since his return from Leipzig. So does that mean we
must assume that Majorana abandoned theoretical physics for four years, only to
take it up again at the end of 1937 when he entered the selection for a university
chair? Some letters to his uncle Quirino do show Majorana going through a "peculiar" period. Indeed, in a letter dated 6 September 1933, when he was just back
from Leipzig, we read:

I have been in Rome for a month. My folks have come in dribs and drabs as well, and will
probably go to Monteporzio in a short time. I will stay in Rome and you should write letters
home as I seldom go to the Institute. I am not going back to Leipzig this year. I believe dad
has stopped doing research. In Rome I have only found Rasetti; Fermi is coming back from
his trip to America in a short while.[44]

And then, in a letter dated 20 February 1935:

I thank you for your wishes regarding my work which I hope will soon be back to a normal
level (that's not saying much).[45]

---

[42]Letter MG/R4 of 27 July 1934 in Recami (1987).

[43]Letter MG/R5 of 20 June 1937 in Recami (1987).

[44]Letter MQ/R4 of 6 September 1933 in Recami (1987).

[45]Letter MQ/R5 of 20 September 1935 in Recami (1987).

In other letters to his uncle Quirino, there is no mention of Ettore's scientific activity, except for a letter dated 16 January 1936, which we will bring up in the next section.

However, for the years we are concerned with here, there is clear evidence that Ettore's "dark period" was not at all a time of abstention from research in physics. A first fascinating part of this evidence is to be found in the letters between Ettore and his uncle Quirino, a noted experimental physicist of the day. They testify to a "collaboration" between the two scientists, even though they did not deal with the kind of frontier research which Ettore had always preferred.

From 1925 to about 1940, Quirino Majorana was involved in experimental research on the phenomenon of photoresistance in thin metal films; he studied the increase in electrical resistance of such films when they are exposed to light under particular conditions. This effect was usually ascribed to the thermal action of light that heats up the film and thus alters its electrical properties. However, Quirino Majorana thought he had identified a new effect which could not be explained through the thermal action of light, but instead through a kind of photoelectric phenomenon differing from the classic effect discovered by Heinrich Rudolf Hertz in 1886. It was the latter whose theoretical interpretation in 1905 had earned Albert Einstein the Nobel Prize (the photoelectric effect refers to the emission of electrons from certain metals when these are illuminated by light of a particular type). Clearly, in order that such an interpretation could even be entertained, it was essential that the thermal origin of the observed effect should be excluded with reasonable certainty; and this could only come out of an appropriate theoretical study, whereupon accurate predictions of the said thermal action could be followed by experiments that would exclude it. Such theoretical calculations were indeed presented by Quirino Majorana in 1938, in what can be considered the conclusive paper on the topic (Majorana Q. 1938). However, a quick comparison between this paper and the contemporary correspondence between Ettore and his uncle[46] clearly shows that the author of the theoretical study was not uncle Quirino at all! It was in fact his nephew Ettore who had provided the appropriate theory for the observed phenomena.[47] Several times uncle Quirino tried, quite rightly, to acknowledge his nephew's work, urging him to publish his theory, or at least let mention his contribution in the 1938 paper; but Ettore, as usual, refused and both the theoretical calculations and the interpretation of the experiments appeared as Quirino Majorana's work.

What may seem like a simple piece of help Ettore gave his uncle to confirm or formulate some mathematical predictions actually hides a deeper and more complex kind of cooperation between the two scientists, one which ignored the considerable age gap between them, but which appears clearly from their private

---

[46]See, in particular, Ettore's letter to Quirino Majorana dated 14 May 1935, kept at the Museum of Physics of the University of Bologna.

[47]An introductory, but particularly interesting, study can be found in (Dragoni 2006). The reference book is (Dragoni 2008).

correspondence. Ettore's interest in his uncle's research dates back at least to the time he was preparing his graduation thesis.[48] At first, Ettore served simply as a "bridge" between his uncle and Fermi; Quirino Majorana really wanted the opinion of the young and brilliant scientist in Rome, and Fermi was quite ready to make suggestions about possible interpretations of the newly observed effect. Such was Ettore's role until his departure for Leipzig. From the end of 1933, Ettore's "remote" involvement in his uncle's research evolved steadily from simple and occasional comments towards active and interested participation. And he did not limit himself to providing the underlying theory for the experiments. Indeed, he suggested technical improvements or identified possible causes of error or secondary effects, so as to control the phenomenon as far as possible.

The collaboration with uncle Quirino extended to other things as well. In this respect it is perhaps significant that Ettore wrote a whole speech his uncle gave as President of the Italian Physical Society (SIF), at the 1937 conference in Bologna to celebrate the 200th anniversary of the birth of Luigi Galvani, attended by many Nobel prize winners and famous scientists from the world over. It is likely that, as the event was approaching, Quirino Majorana (who had had surgery earlier) had asked for help, and his nephew answered right away:

I am so glad to hear that you are recovering. [...] I am only disappointed that I was not so well myself and was not able to visit you during your stay at the hospital. [...]

Dalla Noce has informed me about the conference for Galvani, which I would like to attend, but may have to skip in order to avoid too long a journey from Sicily. I have improvised an opening speech according to your wishes. You may consider it too generic, but it has not been easy to say much more without exceeding the imposed limits. Anyway, if you do not like it, send it back with your comments. [...] Concerning the speech, the set opening and closing are not models of eloquence, but I have included them for their possible links with the rest.[49]

Quirino Majorana's speech turned out to be really a great success, and Ettore made clear his satisfaction: "I have received also *Sapere*. Much admired, and by many, your speech".[50]

---

[48]In this regard, compare the title of one of his oral theses, *Su un effetto fotoelettrico constatato negli "audion"* ("On a photoelectric effect observed in audions") with that of Q. Majorana's article "Su di un fenomeno fotoelettrico constatabile con gli audion" ("On a photoelectric effect observable with audions"), in *Rendiconti della R. Accademia dei Lincei*, vol. 7 (1928), p. 801.

[49]Letter MQ/R28 of 1 September 1937 in Recami (1987).

[50]Letter MQ/R29 of 16 November 1937 in Recami (1987).

Libero docente

On 12 November 1932, Majorana became a lecturer (*libero docente*) in theoretical physics, although it only became official on 21st January of the following year, soon after his departure for Leipzig. Up until now, little importance has been given to this qualification, probably because none of those close to Majorana ever made sufficient mention of it. On the one hand, everyone took it for granted that he would eventually become a professor without any effort, even though we do not know whether he had been encouraged by Fermi, or by someone else, to do so, or whether it was an idea of his own. On the other hand, what happened later on might have led some people to think that Majorana disliked teaching.

The importance to Majorana of becoming a university professor of theoretical physics has only recently come to light, thanks to the unexpected discovery of three documents in the archives of the University of Rome.[51] These documents, undoubtedly signed by Majorana himself, are his official request to give lectures in the University. The first carries the authorisation of the director O.M. Corbino "to use the classrooms and the library of the Institute of Physics for his lessons". Let us examine this in more detail.

The first request dates from May 1933 and concerns the course in *Mathematical methods of quantum mechanics* which Majorana would have liked to give during the academic year 1933–34. It contains the detailed list of the topics to be dealt with in the course, at a rate of three hours a week. It is worth noting that this request was made during the short period Ettore was back in Rome from abroad, from April 12 to May 5, whence it indicates a strong interest on his part, no matter how else one may assess this.

The second request is dated 30 April 1935. The title of the course to be held in the academic year 1935–36 was *Mathematical methods of atomic physics*; here too we find a list of the topics to be dealt with in the course, these differing significantly from those of the previous course.

Finally, the third request was made on 28 April 1936, and concerns the course on quantum electrodynamics to be given in the academic year 1936–37. Obviously, in this case the topics in the scheduled two-hour weekly lessons are totally different from those of the other two courses.

The fact that all three courses were never given is basically certain; not only because his friends and colleagues (Amaldi, Segrè, and others) never mentioned them, but most of all because in the cited documents it is explicitly stated that Majorana "did not deliver courses in the past". Of course, we cannot be completely sure of this as far as the last course is concerned, as we have no subsequent statement. However, it seems perfectly reasonable to extend what has just been said to the third course as well. It is nevertheless extremely interesting that, in the alleged "dark years", Majorana would repeatedly put forward his request to give lectures pretty much every year. The only time he did not was in the year 1934–35

---

[51]This discovery is due to Alberto De Gregorio. See (De Gregorio and Esposito 2006).

for the obvious reason of his father's death, and we know he was very close to his father, or possibly also due to a loss or theft.[52] Though his first request in 1933 was quite possibly made in a period of euphoria due to the experience he was living abroad, where he may well have found working conditions better suited to his personality and abilities than they were in Rome, it is appropriate to remember that he became a lecturer before his stay in Leipzig. The later requests, made while Majorana was living permanently in Rome, thus suggest a sincere, self-determined, and self-motivated interest in teaching, independently of external causes (most likely, a lack of students) which might have prevented the course from taking place.

Another surprising conclusion regarding those years comes from even a superficial analysis of the contents of the courses Majorana had drawn up. These reveal a thorough evaluation of the chosen topics, based on an absolute knowledge and mastery of the contemporary scientific literature, quite different from those dealt with in similar courses (theoretical physics, mathematical physics, and others) of the same period. Different even from those given by Fermi, which were among the most cutting-edge courses in Italy.

There were also significant echoes of these topics in the lessons given later at the University of Naples. The contents of *each* course were drawn up in an amazingly accurate way, given the unprecedented nature of the arguments they dealt with, although they would later become the norm in theoretical physics courses, both in Italy and abroad. This in no way supports the idea that Majorana did no physics at all through the period from 1933 to 1937, just because he no longer frequented his friends and colleagues at the Institute of Physics in Rome. In particular, his second request of 30 April 1935 (right after the gap in the previous year) sounds like a clear reaction to the letter Ettore wrote to his uncle Quirino on February 20, in which he wished for a return to "normal" activity. So, while there is no reason to doubt a succession of one or more "difficult" periods (whose causes we may only speculate about, but which do not seem to be in any way related to his scientific activity), it would nevertheless be an exaggeration to claim that this period lasted for four years, with Majorana shut up in his house as far away as possible from the physics environment. Instead, this seems rather to testify to a change in attitude: he continued to visit the university every now and then (at least to hand in his lecturing requests!), but he no longer took part in the activities of Fermi's group. And this may have a simple explanation if we consider the research being done in Rome during this period, when Fermi and his partners were engaged full time in

---

[52]This last conjecture, which might easily be described as fanciful, has nevertheless one verifiable premise. Anyone interested in this may, for example, take a look at the courses given by Fermi (or others) at the University of Rome, paying a visit to the university archive, where they are still kept. Anyone who does this will not fail to notice the lack of a syllabus for Fermi's earth physics course, given in 1928, lectures that were also followed by Majorana, as we have seen. Such an omission would not be surprising at all if it had been known that the aforementioned document was to be found among Majorana's papers deposited by Amaldi at the *Domus Galileana* in Pisa, as observed by both De Gregorio and myself in 2005. For the moment, we do not have any plausible explanation for this.

fundamental studies of nuclear physics. For while these studies were regarded with respectful interest by the international scientific community, they were of an essentially experimental nature... Majorana may simply have distanced himself from something that did not appeal to his tastes and interests, thus continuing his own theoretical research independently.

Let us end with another alleged "enigma" we hinted at the end of the last section. In a letter to uncle Quirino on 16 January 1936, Ettore wrote: "I have been studying quantum electrodynamics for a while now".[53] In relation to this, it has been noted that "given Ettore's modest way of expressing himself, this means that—and it is of the greatest importance—in the year 1935, Majorana devoted himself deeply to original research in the field of *quantum electrodynamics*. Unfortunately, up to now, there is no trace of his notes, with all the new results he must surely have produced" (Recami 1987). If one studies the 18 *Quaderni* of Majorana's personal notes (Esposito et al. 2008), it is clear that much attention is devoted to several questions of quantum electrodynamics. Unfortunately, in contrast to the *Volumetti* (Esposito et al. 2003), almost none of these *Quaderni* are dated, whence it is difficult to say for sure that his uncle Quirino's letter is referring to the contents of these notes. However, it does seem highly plausible for at least some of the topics Majorana wanted to deal with in his third course. And moreover the choice of such a course was not a matter of chance: it is totally different from the other two courses, and it was chosen prior to 28 April 1936. Once again, it seems to correspond to the letter to his uncle written only three months earlier.

---

[53]Letter MQ/R14 of 16 January 1936 in Recami (1987).

Chapter 3
Ten Short Papers

Tidying up the Spectra

The visible aspect of Majorana's genius, that is, what there was to be appreciated by those of his contemporaries who did not know him well (and such is true until quite recently), is limited to just ten scientific papers, written and published over fewer than ten years.[1] This short list can be found at the end of the book, and it will reveal little to the inexperienced eye, either by the quantity or by the topics, which have nothing particularly "brilliant" about them.

However, Amaldi (1966) was quite disconcerted several years ago when he analysed the first five papers, concerning topics of atomic and molecular physics, in a little less superficial way. They revealed an exceptional quality, both for the deep knowledge of experimental data, down to the slightest detail, and for the ability, most uncommon at the time, to use *symmetry properties* to produce a remarkable simplification of the given problem, as well as providing the most suitable method for quantitative solution.

These first papers fitted roughly speaking into the general framework of activities carried out by the different members of Fermi's group during the years 1928–1932. However, as already stated, only paper P1 was strictly the outcome of the collaboration, within the group, between Majorana and his friend Gentile.

In this paper P1, "On the splitting of the Roentgen and optical terms caused by the spinning electron and on the intensity of the caesium lines", the Thomas-Fermi statistical model is applied to calculate the spectral terms of some complex atoms—such as gadolinium, uranium, and caesium—corresponding to X-ray transitions. The peculiar feature, which immediately leaps to the eye, is that the two authors carry out these calculations by applying the perturbation theory to the Dirac equation, which had only just been discovered (and made known) a few months before. This paper, therefore, is one of the very first applications of the newly born

---

[1]Paper P10 was published posthumously, in 1942, thanks to his friend Giovannino Gentile.

© Springer International Publishing AG 2017
S. Esposito, *Ettore Majorana*, Springer Biographies,
DOI 10.1007/978-3-319-54319-2_3

relativistic quantum mechanics. And it is all the more surprising in that the two authors overcame remarkable difficulties in the numerical calculations, providing a more than adequate quantitative treatment (which included, among other things, some corrections to Fermi's statistical potential), and comparing well with the available experimental data.

The reader expecting something rather more sensational might be disappointed by the lack of "brilliance" of this first paper, and his feelings will probably be confirmed in this by the fact that the numerical values predicted by Gentile and Majorana conform to *present* experiments only within an error of 5%. However, that reader may also be surprised to hear that a better theoretical approximation has (apparently) produced a more accurate numerical result only recently, in 1997 (Lee and Wu 1997).

Majorana's second paper on atomic spectroscopy, paper P3 "On the presumed anomalous terms of helium", which dates to 1931, was inspired by Kruger's (1930) alleged discovery of two new lines in the helium spectrum, and above all by the ensuing debate over their interpretation (were they really lines of helium?). Kruger, as a matter of fact, suggested that these new lines should be interpreted as due to transitions of a helium atom from a "normal" state (with the two electrons in the orbits closest to the nucleus) to a doubly excited level (that is, a state where both electrons are in more distant orbits from the nucleus), but Majorana was not convinced of this. The point was that a process where a helium atom loses one of its two electrons spontaneously—*spontaneous ionisation*—is energetically favoured (and therefore more likely to happen) as compared to Kruger's process with two "excited" electrons.

Majorana's accurate analysis was based on his surprising (and uncommon) mastery of the symmetry properties of the given atomic system, as encoded in Weyl's book on group theory, so well-known to him. This analysis led to the prediction that Kruger's interpretation was valid only for one of the two new lines, while the other could not be attributed to the helium spectrum.

This paper by Majorana (like many others) went largely unnoticed, and so the conclusions here reported were independently rediscovered years later by Wu (1934)—a pupil and collaborator of the famous physicist Samuel Goudsmith, who met Fermi and attended his group in Rome for a while—but also by several others. Oddly, Wu's 1934 paper contained a huge numerical error, and it was only after Wu corrected this mistake (one year later) that his calculations matched Majorana's. Ironically, Wu's paper was much cited in the following years, while no reference was ever made to Majorana's. Other authors who later worked on the Auger effect in helium would also reveal a certain naivety, showing that the problem was not a simple one, so it is no wonder that one can find spectroscopic classifications in these studies that Majorana had already proved to be unrealistic. Only in 1944 did Wu produce new calculations using conditions based on appropriate symmetry properties, thereby permanently reinstating what Majorana had already established. These predictions (from 1931) have been confirmed experimentally to a high degree of precision only in recent years, in fact, in 1994 (Esposito 2014).

Strange as it may seem, however, the most important paper published by Majorana in 1931 was not this one, but paper P5—again on spontaneous ionisation —entitled "Theory of the incomplete P' triplets". The problem was to give an adequate theoretical explanation to the fact that some lines expected in the absorption spectra of atoms such as mercury, cadmium, and zinc had never been observed experimentally. To solve the enigma, which concerned complex atoms with at least two electrons in their outer shells, Majorana introduced a process (in the optical field) which he called spontaneous ionisation, perfectly analogous to the process in the Auger effect described above (applicable in the field of X-ray emission), and today referred to as *autoionisation*. Gregor Wentzel had already considered this process in 1927, and had assumed that the atom spontaneously emitted (at least) one of the two electrons. Majorana followed exactly the same reasoning as Wentzel (1927), but included some substantial changes so that it could be applied to the case at hand. The resulting theory was more complex and complete, and it is no accident that, shortly afterwards (in 1931), Allen G. Shenstone independently introduced the autoionisation process, focussing on the simpler case of the copper spectrum (for which spontaneous ionisation is the rule rather than the exception, in contrast to the case considered by Majorana).

Once again, the ensuing scientific literature referred to Shenstone for the autoionisation process and not to Majorana, but this time the mistake can be attributed to an incomplete understanding of Majorana's complex work rather than to ignorance. In fact, the "bible" of atomic spectroscopy to which everyone has referred since the 1930s, namely, the review published in 1935 by Condon and Shortley (1935), recognised the simultaneous and independent contributions of both Shenstone and Majorana. However, the authors were not fully convinced of the whole scheme proposed by Majorana, because the same autoionisation process should have applied to other cases which no one had yet observed at that time. As Condon and Shortley (1935) put it, "Majorana's theoretical discussion does not show clearly why this should not be so". Actually, it is quite surprising that no spectroscopist at the time was able to identify the "missing" lines correctly (Arimondo et al. 2010), and the question was not solved—from an experimental point of view—until 1955. Once again the experimenters referred only to Condon and Shortley's judgement, and it was not until 1970 that the question was reconsidered in a satisfactory manner from the theoretical point of view, whereupon the precise validity of the scheme presented by Majorana forty years earlier was of course vindicated.

Curiously (although the reader will have come to expect this after what was said in the previous chapter), Majorana did not publish *all* the results he derived for the autoionisation phenomenon, which he described theoretically in the framework quantum mechanics by introducing what he called *quasi-stationary states*. As a matter of fact, in his personal notebooks, we see (Esposito 2014) that Majorana derived many of the important theoretical results published years later (first in 1935, and then in 1961) by Ugo Fano in atomic physics, but also that he was the first to introduce the idea of quasi-stationary states in nuclear physics. And all this was happening well before the 1931 paper, when—at least two years earlier—Majorana was engaged in producing a generalisation of Gamow's theory for his graduation dissertation, in order to describe the nuclear reactions induced by alpha particles in a surprisingly original way (Esposito 2014).

It is clear, and we shall see this in other cases, that the different branches of physics were tightly interwoven for Majorana, and he saw no point in not applying an idea or a method from one area to a different one, if experiments required it.

Molecular Bonds

It was in 1931 that Majorana made most of his journal publications, in fact no fewer than four. Besides the two theoretical studies on atomic spectroscopy already discussed, he wrote two further papers about molecular chemical bonds, and the theory he used there turned out to be useful some time later, when he formulated the theory of light nuclei which brought him international fame among physicists, as we have already seen. As a matter of fact, the explanation for chemical bonds rests precisely on the purely quantum idea of *exchange forces*, that is, forces holding together identical particles—such as the elements of a hydrogen molecule—by the sole virtue of the fact that these cannot be distinguished: if they "swap" place no actual change takes place in the physical system (in the molecule, for instance), and this constrains the two identical particles to stay "together". In 1927 Walter Heitler and Fritz London introduced this idea to explain the stability of the hydrogen molecule H_2 (i.e., its very existence). Two hydrogen atoms H close to each other are bound in a molecule H_2 thanks to the symmetry under exchange of their positions: $H + H \leftrightarrow H_2$. The remarkable agreement between the theoretical predictions and the experimental data confirmed the success of their theory.

A more interesting subject of study than the H_2 molecule (two nuclei with two electrons orbiting around them) was the molecular ion He_2^+, because of the greater number of electrons (three instead of two), making the quantum theory explaining its stability somewhat more complicated, if only due to the different symmetry properties. In his paper P2 "On the formation of the helium molecular ion", Majorana approached this theoretical problem by first examining the symmetry properties of the system. This allowed him to clarify the still confused situation with regard to experiments on the emission spectrum of helium. He then elaborated a generalisation of Heitler and London's theory to explain in detail the formation of the helium molecular ion, the basic idea now being that the exchange takes place between a helium atom He (with two electrons) and its ion He^+ (with just one electron): $He + He^+ \leftrightarrow He_2^+$. Of course, the fact that the exchange now no longer occurs between two identical particles, as in the case of the hydrogen molecule, made everything much more complicated. But Majorana managed to translate this idea into a consistent quantum theory by shrewdly applying the methods of group theory, which he loved so much and, above all knew so well.

The formation of the helium molecular ion was studied almost at the same time (and independently) by Linus Pauling—a future Nobel laureate for Chemistry (in

1954), and for Peace (in 1962)—who in the same year 1931 published a first paper in which he suggested the same idea as Majorana, although he only explored its *qualitative* consequences, without providing a consistent theory. However, Pauling later arrived at the same quantum theoretical results as Majorana (in fact, two years later), but without of course employing the powerful methods of group theory.

Even here, what Majorana reported in his published paper does not completely and faithfully reflect the huge amount of work he carried out on the topic (and, particularly, on helium eigenfunctions). Indeed, he devoted himself to generalising the simple approximations that so often appeared in the scientific literature (Esposito and Naddeo 2012), preserving without exception the physical intelligibility of the resulting expressions. In any case, even the approximations published in paper P2 alone allow us to establish an excellent agreement with experimental observations, and this was not at all "accidental", as Majorana himself put it, in his ever humble and reserved way. Actually, his theoretical predictions went far beyond what could then be observed experimentally: one of the quantities he calculated could not even be compared with any experimental values at the time, although this comparison can now be made rather easily, and we discover that experiments in 1999 (Coman et al. 1999) gave an observed value that was surprisingly closer to Majorana's theoretical estimate of 1931 than the best available theoretical estimate (Ackermann and Hogreve 1991), obtained 60 years later—in 1991—using much more refined mathematical methods than those developed by Majorana.

Paper P4, concerning the "Pseudopolar reaction of hydrogen atoms", is about what appeared to be an even more intriguing scientific mystery. In fact, certain spectral features of the hydrogen molecule had been observed in the infrared region, where they were not at all expected, according to an analogy with the optical spectrum of atomic systems. In particular, the theoretical explanation of the phenomenon (compared also with what happens in atomic systems) had aroused some concern (Weizel 1930). However, this was not at all the case for Majorana. He immediately understood that the situation for molecules was totally different from the atomic case, and immediately gave the required theoretical explanation of the phenomenon.

Majorana realised that, in order to solve the puzzle, the answer was to generalise the theory of Heitler and London. As mentioned above, their description of the H_2 molecule considered only the exchange of two neutral hydrogen atoms: $H + H \leftrightarrow H_2$. Perhaps inspired by his previous paper, Majorana introduced the exchange of two ionised atoms, $H^+ + H^- \leftrightarrow H_2$, and assumed that, in a molecule with a typical "homopolar" bond, there could also be some *ionic structures*. This new idea was quite revolutionary at a time when physicists and chemists kept phenomena involving molecules with homopolar bonds quite separate from those involving molecules with ionic bonds, so Majorana himself tried to "mitigate" its impact by introducing the less evocative term of "pseudopolar reaction". Indeed, the term "ionic structures" used above came much later. Although studies on this topic started to appear in 1949 (Coulson and Fischer 1949), it was only with the re-discovery of paper P4 after the 2006 Majorana centennial (Clementi and

Corongiu 2007) that scientists realised that the introduction of ionic structures in homopolar molecules leads to theoretical predictions that are closer to the experimentally observed values. Since then such configurations have been referred to as *Majorana structures* (Corongiu 2007).

Coming back to Majorana's theory, this not only solved the above-mentioned enigma, but the results he obtained from his impressive calculations turned out to be remarkably consistent with the experimental values. However, that was still not enough to alter his characteristically humble and subdued nature:

> This result is even too favourable as, with the method we followed, we could have expected a value considerably smaller than the true one. [...] A quantitative evaluation is difficult, but it is plausible that such an approximation tends to produce errors compatible with the discrepancies ascertained between calculation and experiment.

A Formula and a Multifaceted Theorem

The topics Majorana thought were worth studying, he chose himself, often after reading the scientific papers in the well-stocked library of the Institute of Physics in Rome. But sometimes, as already noted, it was one of Majorana's friends and colleagues who aroused his interest, and the most significant case was certainly that of the work leading to paper P6 about "Oriented atoms in a variable magnetic field". However, while the article came into being through the request to solve a specific problem, the achievement of the solution cannot accurately convey the idea of the content—both specific and general—of the article, as it goes well beyond expectations (and not only those of his friend Segrè).

The central topic is the study of the dynamic behaviour of a spin-½ particle in a region where there is a time-varying magnetic field that almost vanishes at a given instant, as required by Frisch and Segrè's experiment. As Amaldi recalls, this study was carried out "with extraordinary elegance and conciseness", but it is in the method Majorana followed to solve the problem that the most interesting results are to be found. *Unusually* as compared with the procedures adopted by his fellow physicists, he would first solve a more *general* problem—concerning the dynamics of a particle with arbitrary angular momentum—and then *reduce* this to the specific case under investigation, namely, a spin-½ particle.

Of course, the "extraordinary elegance and conciseness" of Majorana's solution for the general problem was achieved using his favourite methods of group theory. As the reader should be aware by now, this meant that it would only be recognised some years later. In this case, it happened in 1937 thanks to Isidor Isaac Rabi (and, in more general form, in 1945, thanks to Felix Bloch and Rabi, both future Nobel laureates): but it was of course done using a less difficult (for the physicists of the time) graphical representation, introduced this time by Majorana himself. This representation is known to mathematicians[2] as the *Majorana sphere*, but it is sometimes still improperly referred to as the *extended Bloch sphere*. In his paper, Bloch considered only one particular case (which, not surprisingly, was precisely the one Frisch and Segrè needed), so the general case—if one ignores Majorana's work of more than ten years before—is referred to as *extended*... The utility of the representation in terms of this "sphere" comes mainly from the fact that the action of a variable magnetic field simply "rotates" the points on the sphere, and its "conciseness" is the reason for the success of this representation, which is to be found in so many quantum mechanics textbooks.

However, as mentioned earlier, the most (apparently) "abstract" part of Majorana's theory was not immediately appreciated, despite Bloch and Rabi recognizing that Majorana's paper "has given some general results which are both very useful and greatly deepen our understanding of the process involved". The language and powerful methods of group theory were still particularly difficult in the eyes of physicists in those days, and even in 1945, Bloch and Rabi felt the "obligation" to provide an adequate *vulgata*:

> Majorana's method, while remarkable in its elegance, has the disadvantage of somewhat obscuring the physical significance of the representative systems with spin ½. It is clear that a simple intuitive understanding of the procedure and the essential formulae will be very useful to many (Bloch and Rabi 1945).

The true essence of the more general results obtained by Majorana was realised some time later by another Nobel laureate, Julian Schwinger, although unfortunately, he did not publish anything on the matter for forty years, as he himself points out:

> At the heart of the attitude that I adopted toward angular momentum theory was the celebrated theorem and formula of Majorana. This established, qualitatively and quantitatively, how the behaviour of an arbitrary magnetic moment in a time-varying magnetic field is related to that of a spin-½ system. The original Majorana paper was baffling and it was obligatory to find a quantum mechanical derivation of the formula, in particular. My answer was only hinted at in a 1937 paper (Schwinger 1937) [...]. The subject remained quiescent until 1945 (the year for renewed attention to basic physics) when Bloch and Rabi remarked on the derivation of Majorana's formula from the spin-½ representation, and I began to write a paper supporting the thesis that the expression of symmetry concepts in quantum mechanics does not require the injection of group theory as an independent mathematical discipline. A major piece of evidence was to be the derivation of Majorana's theorem and formula, thereby finally making explicit the veiled reference of 1937. For some reason this paper was never completed (Schwinger 1977).

Terms such as *Majorana's formula* and *Majorana's theorem* in the acknowledgement reported above appeared only in 1977, but the full weight of their meaning is by now well recognised.[3]

---

[2]It is in fact totally general, and allows the graphical representation of a state with arbitrary angular momentum.

[3]This is also attested, for instance, by the fact that the Schwinger paper (1977) we were quoting from was included among the seminal papers of the Nobel laureate (Milton 2000).

In recent times, the mathematician Penrose (2000) has rediscovered the general result of the *Majorana representation* (without Bloch's "reduction"), providing a geometrical interpretation which has proved to be particularly useful in quantum information science: amazingly then, Majorana's name has even entered the world of *quantum computing...* Many other mathematical applications of the Majorana representation have appeared in the literature in recent years (probably thanks to Penrose's very popular papers), but it would be pointless to go into the details here. For as we were saying, the mathematical aspects of paper P6 are just one side of the whole story, whose purely physical implications are equally interesting.

First of all, it is appropriate to point out that a similar problem to the one studied by Majorana was considered almost at the same time and independently by Landau (1932), Zener (1932), and Stückelberg (1932), who obtained the same results as Majorana (regarding the so-called probability of spin-flipping, sometimes known as the *Landau-Zener probability*). However, they employed more sophisticated mathematical methods (especially in Stückelberg's case). This easily explains why, despite the fact that the formula for the Landau-Zener probability is used in very different contexts, from atomic and molecular applications to the phenomenology of neutrino oscillations, Majorana's easier and more transparent approach is now commonly adopted in quantum mechanics textbooks, although of course they almost always overlook the reference to Majorana.

The physical applications of Majorana's work are many, starting from the first remarkable example already cited and mentioned in Bloch and Rabi's paper, which laid the theoretical foundations for the experimental methods of nuclear magnetic resonance (NMR).

In more recent times, however, paper P6 has also played a fundamental role in the realisation of a remarkable physical phenomenon, the *Bose-Einstein condensation* of atomic gases, as a result of which Eric A. Cornell, Carl E. Wieman, and Wolfgang Ketterle shared the Nobel Prize in 2001. Such a phenomenon can be observed only by cooling the atoms down to very low temperatures (close to absolute zero, i.e., about minus 273 °C), and this can be achieved using "magnetic traps". When atoms move in such traps, they are confined by a variable magnetic field, and the more slowly they move around the local minimum of the magnetic field, the colder they get. So it is easy to see that such a situation is perfectly similar to the one studied by Majorana, where the atoms move close to a region in which the magnetic field vanishes—and which is now referred to as *Majorana hole*. However, an unwanted effect can arise, as the Nobel laureates themselves explain:

> When constructing a trap, for weak-field seeking atoms, with the aim of confining the atoms to a spatial size much smaller than the size of the magnets, one would like to use linear gradients. In that case, however, one is confronted with the problem of the minimum in the magnitude of the magnetic fields (and thus of the confining potential) occurring at a local zero in the magnetic field. This zero represents a "hole" in the trap: a site at which atoms can undergo Majorana transitions and thus escape from the trap. [...] We knew that once the atoms became cold enough they would leak out the "hole" in the bottom of the trap (Cornell and Wieman 2002).

An estimate of the *Majorana flop rate* was then obtained, to predict how to find a solution to the problem, and hence observe the desired phenomenon of Bose-Einstein condensation.

One last application (among many) that is worth noting is surely the phenomenon known in spectroscopy as the *Majorana-Brossel effect*. From the beginning of the 1950s, some experiments were carried out to investigate atomic structures in which the atoms were doubly excited (that is, two electrons had an energy above their normal level), these states being induced by some particular incident electromagnetic radiation. While Bloch and Rabi's papers had previously focussed on nuclear magnetic resonance induced in atomic beams, other scientists were now trying to demonstrate magnetic resonance in solids and liquids (besides gases), using optical detection methods that had been improved mainly by Jean Brossel. The fundamental experiment to this end was done by Brossel and Bitter (1952) who studied the polarisation of the fluorescence emitted by mercury. The effect they observed was interpreted in terms of Majorana transitions, and the formula for the magnetic resonance transition probability in paper P6 was once again employed to give a correct explanation of the experimental data.

> As is apparent from the form of the Majorana expression for the transition probabilities, when more than two equally spaced levels are involved in a resonance, various multiples and submultiples of the normal Larmor frequency may be expected. The modulation of the resonance radiation emitted by a processing atom is formally equivalent to a "beat" phenomenon between two coherent radiations whose frequencies differ by the amount of the processional frequency (Bitter 1962).

Generally today, we refer to a Majorana-Brossel effect whenever a non-trivial alteration of the shape of the spectral lines is induced by strong radiofrequency magnetic fields. The topic is still under investigation, and another Nobel laureate—Claude Cohen-Tannoudji—has recently re-interpreted such phenomenon in a more general way, by introducing a process he called *Majorana inversion* (Cohen-Tannoudji 2003). We thus see that the saga of the Majorana "terms" suggested in paper P6 is destined to continue...

Heisenberg Does Some Promotion (But not Too Much)

Before Chadwick's discovery of the neutron in 1932, the commonly accepted model for nuclear structure considered an atomic nucleus made up of protons and electrons—the only particles known at the time—, which were emitted during particular reactions involving radioactive nuclei. In a series of three papers (Heisenberg 1932a, 1932b, 1933) written between June and December 1932, Heisenberg reformulated the theory by introducing the neutron (together with the proton) as a constituent element of atomic nuclei, showing that there was apparently no need to hypothesise the existence of nuclear electrons. As pointed out in the previous chapter, the forces responsible for "holding together" protons and neutrons were of the same nature (*exchange forces*) as the ones introduced by Heitler and London for molecules, something Majorana had considered in detail in his 1931 papers. But Heisenberg himself was wavering curiously between an idea of the neutron as an "elementary" particle and one in which it was made up of a proton and an electron (Heisenberg 1933). The latter position was strongly suggested by the emission of electrons in the radioactive processes known as *beta decays*. Despite the undeniable and inescapable improvement brought about by the proton + neutron nuclear model compared with the proton + electron model, such hesitation was typical among physicists at the time. Indeed, it would be some time before the Heisenberg model was completely accepted. But the situation totally changed with the appearance, in 1933, of paper P8 by Majorana—whose origins we have already discussed—and the publication, almost a year later, of "Fermi's theory of beta decay" (Fermi 1934b), which showed without any doubt that the electrons emitted in these decays did not pre-exist in the nucleus.

To understand why Majorana's paper P8 "On nuclear theory" managed so successfully to change the traditional view, and hence to fully appreciate Heisenberg's "promotion" of it, to the point of bringing lasting fame to the physicist from Sicily, one has simply to follow Majorana's own arguments, starting from a clear analysis of Heisenberg's work.[4]

> In the absence of other guiding criteria, Heisenberg was guided by an analogy presumably existing between the normal neutral hydrogen atom and the neutron, the latter being supposed – as commonly accepted – to be composed of a proton and an electron. [...] The use of such an analogy is difficult to justify since, if the neutron were effectively composed of a proton and an electron, their binding could not be described by present theories, which would lead to one associate the Bose-Einstein statistics and an integer multiple of $h/2\pi$ for the mechanical moment to the neutron, contrary to fundamental assumptions. These come directly from empirical properties of the nuclei, and we cannot give them up. Given the present state of our knowledge, it is thus preferable to try to obtain the law of interaction between the elementary particles being guided by simplicity alone, in order to predict the most general and characteristic properties of nuclei.

Here the key word characterizing Majorana's work is "*simplicity*", and it emerges as soon as one compares his theory with the abstract formalism introduced by Heisenberg, for instance, when the problem is to describe a nucleus made up of "impenetrable" matter (so not like atoms). The basic hypotheses of the new theory are exactly those adopted by Heisenberg, but now there is no hesitation over the *elementary* constituents of the nucleus, so there is no risk of falling into the same traps as the German physicist. Quite the contrary, what Majorana means by *simplicity* is clear when he describes the forces holding together protons and neutrons in the nucleus. Although he uses a *simpler* mathematical formalism, Majorana adopts the same *exchange* mechanism as introduced by Heisenberg, but *exactly* as it

---

[4]Majorana published his paper P8 both in German and in Italian (about three months later), with *almost* no difference between the two versions. The most relevant difference (in my opinion) may lie in the fact that the author is a little more explicit in his criticism of Heisenberg's theory in the Italian edition. In the following quotes we will refer to this version.

appears in Heitler and London's molecular theory: protons and neutrons are bound by the fact that, swapping positions, there is no actual change inside the nucleus, just as in the hydrogen molecule model. This swapping of positions, and *not* electrical charge as in Heisenberg's model, led straight to phenomenological predictions in agreement with experimental observations, and this immediately convinced Heisenberg of the superiority of Majorana's theory.

In stark contrast to his other papers, the ensuing exposure of this one meant that the scientific community quickly adopted Majorana's exchange mechanism. This is easy to see, not only in the scientific papers of the 1930s, but also in subsequent reviews and books, such as the well-known textbook by Blatt and Weisskopf (1952). But surprisingly, paper P8 contains two other important contributions (as always, not at all emphasized by its author), which were only rediscovered some time later by nuclear physicists, without any reference to Majorana, probably because Heisenberg did not immediately realise their relevance in this case.

The first contribution concerns the application of the Thomas-Fermi model to nuclei, rather than atoms; here of course Majorana had a running start, thanks to his experience with the statistical model he had acquired previously. With this application he was able to demonstrate that his model predicted a nuclear energy minimum corresponding to a stability condition for the nucleus (the nucleus does not disintegrate under such conditions), in full analogy with what happens in the Heitler-London molecular case. The Thomas-Fermi model was only applied to nuclei from the 1960s (Seyler and Blanchard 1961), thus revealing Majorana's foresight once again.

The second contribution deals with the explicit form of the interaction between protons and neutrons, which did not actually enter directly into the main calculations carried out by Majorana, or even Heisenberg. Anyway, Majorana did consider this topic too, however fleetingly, realising what Yukawa (1935) discovered independently two years later, which is referred to today as the *Yukawa potential*. Majorana did not elaborate in detail on this intuition, while Yukawa explored the physical consequences of the form adopted for the interaction potential, by introducing the so-called *meson theory*. The reason for this approach is easily ascribed to purely practical considerations, something Majorana was not in the habit of avoiding: "we shall not follow this up, since it has been shown that the first statistical approximation can lead to considerable errors, however large the number of particles". His hypercritical judgement had once again hit the nail on the head, as the future would show.

Infinite Problems for Professor Pauli

The actual birth of theoretical elementary particle physics, as we understand it today, can be dated back (Giannetto 1993) to paper P7 entitled "Relativistic theory of particles with arbitrary intrinsic angular momentum", published by Majorana in 1932 (and in some respects, to paper P9, published in 1937). The reason lies mainly in Majorana's conception of relativistic quantum field theory, and the way he exploited it in his work (some published but mostly unpublished), and all this way ahead of his time.

The birth of quantum mechanics in the 1920s was built upon the basic principles of Einstein's relativity, so it is not surprising that its founding father Erwin Schrödinger *first* wrote the relativistic wave equation (which would later be referred to as the Klein-Gordon equation), and *only then* his more famous non-relativistic equation. The dismissal by Schrödinger (and others) of the relativistic Klein-Gordon equation as the correct quantum equation describing the electron was based decisively on the comparison between its theoretical predictions and accurate spectroscopic data obtained experimentally for the hydrogen atom: the discrepancy was at once correctly attributed to the fact that the electron spin had not been taken into account in the theoretical framework. Oddly enough, when Dirac (1928a, 1928b) developed the appropriate formalism leading to his well-known relativistic equation (which resulted in a very good agreement with spectroscopic data), he did not pay much attention to such discrepancies, but focussed rather on a different *theoretical* problem, namely, how to obtain a theory which would attribute *positive* probabilities for localising an electron inside an atom. Indeed, this was the case for Schrödinger's equation, but not for the Klein-Gordon equation. The second theoretical problem which came up, that is, the presence of electron states with negative energy (clearly without any physical significance), was initially left unsolved by Dirac. However, Dirac himself (1931) later provided an interpretation of his *holes* in terms of antiparticles with the same mass as electrons, but opposite electric charge.

With the exception of Pauli[5] and a few others who (quite rightly) ascribed importance to the theoretical problem of the negative energy states, the majority of theoretical physicists soon recognised how successful Dirac's theory was, even before its ultimate confirmation with the discovery of the first antiparticle, the positron (Anderson 1932; Blackett and Occhialini 1933). Indeed, the excellent experimental agreement with the theoretical predictions of hydrogen fine structure and the electron magnetic moment soon convinced everybody of the validity of the Dirac equation. Dirac's theory offered a sensible and attractive solution to the main problems of the newly born quantum mechanics, which were closely related to the electron spin, and furthermore, the theory was founded on a solid mathematical formalism. This convinced many that the value of ½ for the electron spin was a *necessary* consequence of the theory of relativity, when applied to the basic principles of quantum mechanics (Gentile 1940).

But Majorana's 1932 paper P7 demonstrated exactly the opposite: a relativistic quantum theory could be built for particles with arbitrary spin, provided that the persistent problem of negative energy states was subjected to a lucid examination.

---

[5]"I do not believe in your perception of 'holes', even if the existence of the 'antielectron' is proved", Pauli would write to Dirac on 1 May 1933 (Pauli 1985).

The Dirac equation was a favourite topic, and studied at length (Esposito et al. 2003, 2008) by Majorana, who (perhaps unexpectedly) took it as a formal starting point for his theory. As a matter of fact, Majorana's equation has exactly the same form as Dirac's, but his analysis of the problem of the negative energy states led him to the conclusion that the wave function describing the given particle should have an infinite number of components, in contrast to the Dirac theory, which considered only 4 components. The difficult mathematical problem Majorana had to solve, within the framework of group theory, is known in the technical jargon as the problem of infinite-dimensional unitary representations of the Lorentz group, something that would only be dealt with years later by the mathematician Wigner (1939, 1948). The latter knew Majorana's 1932 paper well, but curiously enough, considered that it was not very rigorous…

> The representation of the Lorentz group has been investigated repeatedly. The first investigation is due to Majorana, who in fact found all representations of the class to be dealt with in the present work excepting two sets of representations. […] The difference between the present paper and that of Majorana […] lies – apart from the finding of new representations – mainly in its greater mathematical rigor. Majorana […] freely uses the notion of infinitesimal operators and a set of functions to all members of which every infinitesimal operator can be applied. This procedure cannot be mathematically justified at present, and no such assumption will be used in the present paper. Also the conditions of reducibility and irreducibility could be, in general, somewhat more complicated than assumed by Majorana (Wigner 1939).

Wigner's greater mathematical rigour certainly accounted for the 56 pages of his paper, but the answer given in advance by the physicist Majorana in his paper P7 is particularly enlightening:

> In order to avoid exaggerated complications, we will give the transformation law only for infinitesimal Lorentz transformations, since any finite transformation can be obtained by integration of the former.

Such an approach is just what almost all theoretical physicists actually adopt today.

Despite a certain simplicity in the reasoning, Majorana's theory was not immediately understood by many. While this pioneering work was barely noticed at the time (Fradkin 1966) by the international scientific community (perhaps because it was published in an Italian journal, little known abroad), it was also largely misunderstood, being too far ahead of its time. This was the case for Fermi himself (Recami 1987): when describing Majorana's scientific works for a university selection, he overlooked precisely this important paper.

First of all, there was the conviction, as mentioned above, that spin-½ particles must exhaust all of Nature's creativity, and that this resulted directly from the theory of relativity through Dirac's simple equation. It was only in 1934 that Pauli and Weisskopf showed that this was not true, and that a quantum theory for particles with different spins (specifically, with 0 spin) might be built without any conceptual problem, predicting physical processes perfectly similar to the ones expected in Dirac's theory for spin-½ particles. As Weisskopf remembers, there was tremendous fun in working out something that, at that time, was quite unexpected: that one can get pair creation and annihilation without a Dirac equation, also for particles without spin (Weisskopf 1973).

Actually, for Majorana there was nothing unexpected about this. Some years earlier, he had already elaborated (without publishing) the theory that would later be known as *Pauli-Weisskopf scalar electrodynamics*. In particular, an amusing episode occurred during a lunch break at the International Conference of Nuclear Physics organised in Rome in October 1931 by Fermi and Corbino. It is recalled here by the young Gian Carlo Wick (together with Majorana and Heitler):[6]

> I was asked by Heitler to act as a sort of interpreter between him and Majorana. He spoke hardly any Italian, and Majorana's German was a bit weak. So during lunches, Heitler expressed a curiosity in what Majorana was doing; Fermi must have told him how bright he was. So Majorana began telling, in that detached and somewhat ironical tone, which was typical of him, especially when discussing his own work, that he was developing a relativistic theory for charged particles. It was not true, he said, that the Schrödinger equation for a relativistic particle had to have the form indicated by Dirac. It was clear by now, that in a relativistic theory one had to start from a field theory and for this purpose the Klein–Gordon wave-equation was just as legitimate as Dirac's. If one took that, and quantized it, one got a theory with consequences quite similar to Dirac positron theory, with positive and negative charges, the possibility of pair creation, etc. You can see what I am driving at: Majorana had the Pauli–Weisskopf scheme all worked out already at that time. [...] Heitler probably forgot all about it, and so did I, until I saw the paper by Pauli and Weisskopf.

> I must confess that I knew enough to understand the gist of what he was saying, but not enough to appreciate how novel and original it was. Heitler probably did, because his comment, as I recall it, was: "I hope you will publish this." [...] [But] in the case I have described publication never occurred (Wick 1981).

What Wick and Heitler witnessed is diligently reported in Majorana's *Quaderni* (Esposito et al. 2008), and the whole theory (with applications, too) has been carefully reconstructed in recent years (Esposito 2007b), showing how fully Majorana had understood the situation—since 1929–1930.

Majorana's theory with the infinite-component equation also had some particular physical consequences which were rediscovered and appreciated only years later (Esposito 2012). First of all, in contrast to Dirac's equation describing the quantum behaviour of just one spin-½ particle (or, better, one particle/antiparticle pair), the Majorana equation was a multi-mass equation, which described at the same time different particles with different spins, each with a specific mass predicted by the theory. Furthermore, as noticed immediately by Majorana, certain solutions to the equation corresponded to mass values of particles moving with a speed *faster* than the speed of light. For the first time, tachyonic solutions entered a relativistic wave equation.

---

[6]This episode is also quoted in (Recami 1987) and, above all, in an unpublished note by Wick, kept in the Wick Archive at the Library of the *Scuola Normale Superiore* in Pisa. The theoretical calculations we are speaking about here are reported in Majorana's *Quaderni*, kept in Pisa.

Certainly, all these peculiar characteristics—and many others in unpublished "preparatory" works (Esposito 2012)—did nothing to promote the understanding of such a very advanced paper. An emblematic case was Pauli, whose highly developed theoretical acumen is (and was) well known. Probably informed by Heisenberg about Majorana's paper P7, Pauli studied it at length (at least from 1939 to 1947) over two separate periods, with his collaborator Marcus Fierz (Pauli 1993).[7] The reason for such a prolonged interest is given by Pauli himself in a letter to the theoretical physicist Homi J. Bhabha on April 12, 1940:

> I believe in the existence of many more particles than are yet known, particularly particles with arbitrary values of the spin and charge.
>
> My considerations of particles with higher spins is coming to some end now. I think that they exist really, but I can fancy that the complication of the theory comes from the assumptions, that one has to describe a set of particles with a finite number of spin values only. Maybe the matter becomes simpler, if one introduces a priori an infinite set of spin values (compare Majorana […]).

Pauli's suggestion was, of course, promptly accepted by the Indian physicist who, in 1945, achieved results (Bhabha 1945) already partially obtained (but not published) by Majorana (Esposito 2012). But for a long time, Pauli still did not fully understand paper P7:

> Anyway, it seems to me that the case of the infinite-dimensional representations of the Lorentz group has not been studied sufficiently. [..] Read Majorana! (Pauli to Fierz, July 3, 1940).

And this is true for the mathematician Wigner as well, despite what he wrote in his paper from 1939 (as recalled above):

> Wigner did not understand Majorana's equations, as he admitted to me (Pauli to Fierz, March 29, 1941).

Although Pauli recognised the problem of tachyonic solutions to Majorana's equation as "pathological", he continued to study it for a long time, and also obtained some relevant mathematical results; and the same goes for his collaborator Fierz, up until 1947. But the problem Majorana had faced was clearly too difficult, and would have to wait for later developments.

These came some 20 years later. The revival of Majorana's pioneering paper can be attributed to the "review" by Fradkin in 1966: it was then recognised that Majorana's work provided a general method for developing a whole series of theories of elementary particles, which would become the model from the 1960s on.

The first result was the publication of an English translation of the paper P7 in 1968 by Claudio A. Orzalesi for the University of Maryland. This clearly encouraged new investigations of Majorana's infinite-component equation, and these were quick to come. While only a couple of articles appeared quoting Majorana's work between 1932 and 1960, about 70 appeared in the literature

---

[7]The following quotations from letters are taken from this bibliographic reference.

between 1966 and 1970, dealing with various problems relating to Majorana's theory (Esposito 2014). In 1971, Dirac himself attempted a curious theory, in a sense halfway between his own and Majorana's.

> One of the surprising different physical consequences of the new theory is that it allows only positive energy values for the particle. A relativistic wave equation allowing only positive energies was proposed a long time ago by Majorana (1932). There is a connexion between Majorana's equation and the present theory. [...] It would just be Majorana's, in a different representation. [...] We shall here keep to the whole set of equations of part I and so have only one mass value (Dirac 1972).

The physical interpretation of the "pathological" problem noted by Pauli came only in 1973 (Barut and Duru 1973), and this later led to the conclusion that "the existence of [tachyonic] solutions of infinite-component wave equations is not a 'disease' but a virtue", since they "have a definite physical interpretation, they have experimentally demonstrable consequences and form an integral part of the theory" (Barut and Nagel 1977).

The interest in Majorana's paper P7 has kept up almost continuously until today, mainly because of the popularity of "string theory". As observed recently, it anticipated "to some extent, both Regge's idea and its eventual realization in the Veneziano amplitude, and thus in string theory, by over thirty years!" (Sagnotti 2013).

Pulling the Neutrino Out of a Hat

Among scientists (but not only), Majorana's name is *today* universally associated, roughly speaking, with his paper P9 on a "Symmetric theory of electrons and positrons", mainly because of its predictions about what are commonly known as *Majorana neutrinos* and *Majorana fermions*. Now it is particularly interesting to see that, when Majorana elaborated the theory that would later prove to be fundamental for the physics of elementary particles, he did not want to divulge his discovery at all. In fact, while it is true that it was published in 1937 (perhaps encouraged by Fermi at the time of the selection for the tenure, as we will see in the next chapter), it is also true that it had already been elaborated some time before: we find traces of it in the *Quaderni* and even in the syllabus of the optional course on quantum electrodynamics (of 28 April 1936), which explicitly mentions "the positive electron and the *symmetry* of charges".

As for the theory of paper P7, here again Majorana's main purpose was not to write a different equation from Dirac, but rather to reformulate the already existing equation in order to achieve a complete "symmetry" between the electron and positron components it described. Although in 1933 Heisenberg had noticed[8] this substantial symmetry in Dirac theory, among processes involving electrons and

---

[8]See Heisenberg's letter to Sommerfeld of 17 June 1933, reported in (Pauli 1985).

positrons,[9] the general idea of particle/antiparticle symmetry was formally developed by Majorana in his paper P9. And the target was reached in a surprisingly easy way: once again, the proposed equation has *exactly* the same form as Dirac's, but Majorana's choice of the four Dirac matrices appearing in it differs from Dirac's. In one fell swoop, the electron equation becomes *identical* with the positron equation (in contrast to what happens in Dirac's theory), and generally speaking the quantum description of the electrically charged particles is totally symmetric with respect to the particle and antiparticle states. Majorana, of course, was well aware that such an advantage over Dirac's theory was essentially formal: there was no distinction between the two theories as regards physical applications (but with the remarkable result, emphasized by Majorana himself, that the cancellation of infinite-valued constants is now *required* by the symmetrisation of the theory). Anyway—and this is the great innovation—Majorana's equation admits one more solution than Dirac's, and *different* from his, which represents an electrically neutral particle. In Majorana's words, the new theory represents "the simplest theoretical description of neutral particles", where there is no need to introduce antiparticles.

> [Our approach] allows not only to cast the electron–positron theory into a symmetric form, but also to construct an essentially new theory for particles not endowed with an electric charge (neutrons and the hypothetical neutrinos). Even though it is perhaps not yet possible to ask experiments to decide between the new theory and a simple extension of the Dirac equations to neutral particles, one should keep in mind that the new theory introduces a smaller number of hypothetical entities, in this yet unexplored field.

Majorana would not settle for an intuitive result, so to speak, but wanted to provide solid theoretical grounds, from a *variational principle*, as appropriate to the formalism of quantum field theory. As shown earlier by Dirac and Pauli–Weisskopf, this was not a trivial task, but Majorana, guided purely by mathematical elegance and symmetry, succeeded in bringing theoretical respectability to the idea that spin-½ particles could be their own antiparticles, i.e., he made it consistent with the general principles of relativity and quantum theory.

Such a result was recognised by many[10] as soon as P9 first appeared, including the rather sceptical Pauli who appreciated "the procedure of Majorana" (Fierz and Pauli 1939; Pauli 1941). In particular, just a year later, Furry (1938) was the first to recognise "Majorana's interesting new type of variation principle, concerning which we have nothing new to add", and to firmly and lastingly link that theory to the neutrino.

In fact, there was something new to add, arising from the physical consequences of the new theory, but these came only slowly as time went by, and once again, Majorana had proved to be too far ahead of his time.

---

[9]A very similar topic is dealt with in (Heisenberg 1931), although focusing more on applications than purely theoretical: the symmetry between holes and electrons in an occupied atomic level or in an occupied energy band of a crystal.

[10]Among the first to be impressed by Majorana's theory and its consequences, we may mention Giulio Racah, Hans A. Kramers, Wendell Furry, Nicholas Kemmer, Eugene Wigner, Frederik J. Belinfante, and, of course, Wolfgang Pauli (Esposito 2014).

The very possibility that a particle (of spin-½) might be its own antiparticle—undoubtedly revolutionary for the time, because directly opposed to what Dirac had assumed a few years earlier, and which had been confirmed by the discovery of the positron—was not immediately accepted, and the first obvious question was: are there any candidates for this? With unprecedented foresight, Majorana had already suggested in paper P9 that the *neutrino*, whose existence had just been proposed by Pauli and Fermi to explain the puzzling features of nuclear beta decay, might have just these features.

The first to study the consequences of Majorana's hypothesis, a few months after the publication of P9, was a visitor to Fermi's group in Rome, Giulio Racah, who may well have been informed by Pauli about the new theory (Esposito 2014). Anyway, the most appropriate suggestion for testing Majorana's hypothesis was proposed only a year later: if a neutrino and antineutrino were to coincide, then a new nuclear process—the *neutrinoless double beta decay*—could take place in Nature, something which could not happen otherwise. Over the course of time, the search for such a rare process has proved to be extremely difficult, but given that this question is of crucial importance for elementary particle physics (and not only), it is still being actively pursued in laboratories around the world.

It is also interesting to see that another hot topic in present neutrino physics research—*neutrino oscillations*—made its first appearance thanks to Majorana's work (although indirectly), and thanks to another member of Fermi's group, who had since emigrated to the USSR. In 1957, Bruno Pontecorvo took the step of assuming that neutrinos were particles with the features conceived by Majorana (Pontecorvo 1958a, b). Here experimental research has been more successful than in the previous case, and in 1998 (Fukuda et al. 1998) we obtained the ultimate confirmation that the neutrino oscillation process does actually take place in Nature (even if according to a variation of the original idea suggested by Pontecorvo); but experimental and theoretical research is still continuing all over the world to grasp the key features of this elusive particle, features which have consequences in various branches of physics, astrophysics, and cosmology.

Apart from the neutrino, other particles with the properties conjectured by Majorana (*Majorana particles* and *Majorana fermions*) invaded another branch of theoretical physics in the 1970s, when the theory of *supersymmetry* (or, better, *supersymmetric theories*) were developed. Arising in a sense as a quantum-mechanical enhancement of the properties and symmetries of space-time, the idea of supersymmetry has proved to be important to explain various theoretical aspects of the physics of elementary particles (and also cosmology). Fundamental ingredients of the supersymmetric theories are a plethora of new particles, all of which necessarily display the features assumed by Majorana, and along with neutrinos, this is the reason for the prodigious success of Majorana's theory in paper P9. Tens of thousands of (theoretical) papers published over a few decades have spread terms like *Majorana fermions* or *Majorana spinors* everywhere. And this has happened despite the fact that laboratory experiments have yet to confirm the evocative predictions of supersymmetric theories.

Even more intriguing is the fact that the clearly extraordinary new idea introduced by Majorana's paper P9, which one might say was typical of elementary particle physics, has naturally migrated to the completely different field of condensed matter physics. The first appearance of Majorana fermions in this new area came towards the end of the 1980s (Boyanovsky 1989; Tsvelik 1990) in the study of the so-called ferromagnetic chains (a ferromagnet is nothing other than a magnet), and continued in the following years for similar systems (Esposito 2014). Anyway, the real "revolution" began with the systematic study of the so-called Majorana zero-modes arising in solid-state systems: these have precisely the property of being their own antiparticles. Everything started in 2000 (Read and Green 2000) when the concept introduced by Majorana was applied to fractional quantum Hall liquids, but it is still continuing now in many other physical systems that would be impossible to list completely here. The reason for such an incalculable number of applications lies in a rather particular property ("non-Abelian exchange statistics") of the Majorana zero-modes which will allow us to use them as the building blocks for *quantum computers*, the computers of the future (Wilczek 2009).

Other surprises probably await us in the years to come, given the extraordinary wealth of paper P9.

Something Unexpected

In addition to the nine papers discussed above, it is now customary to add another, published posthumously (that is, after Majorana's disappearance) in 1942, and edited by his friend Giovannino Gentile when he was asked to write a biographical article for the *Enciclopedia Italiana*. Its inclusion in the list of works published by Majorana is fully justified by the fact that the article was considered "complete" and ready to be published by the author himself, who wanted to present a physicist's point of view on "The value of statistical laws in physics and social sciences". As Gentile remembers in his introduction to paper P10, "this article has been conserved by the dedicated care of his brother and it is presented here, not only for the intrinsic interest of the topic, but above all because it shows us one aspect of the rich personality of Majorana, which so much impressed people who knew him; a thinker with a sharp sense of reality and an extremely critical, yet not sceptical mind".

The genesis of paper P10 is particularly interesting, given the somewhat unusual topic it deals with, and reveals what may be described as a feeling of "family duty" on the part of Majorana, something we are already familiar with.

By the end of 1935, his uncle Giuseppe Majorana retired from his post as professor of general economics at the University of Catania, and given his particularly brilliant academic achievements, the Faculty of Law decided to publish a volume in honour of the famous scholar, with contributions from eminent personalities of the day. Giuseppe Majorana himself organised the structure of the volume, and in January 1936 asked his nephew Ettore if he would make a personal contribution as an eminent "mathematician":

> Mathematics has many points of contact– methodological, investigative, etc. – as well as dissonances with social disciplines. The Faculty [of Law] in Catania will publish a book in my honour for my retirement. Could you send me a paper relating to your own fields of interest? One of our common interests is that of the "statistics" of infinitesimal corpuscles, upon which we talked about previously. Could you, e.g., outline the theoretical framework of such an investigation, perhaps comparing statistics as otherwise (or commonly) understood? (Roncoroni 2012).

Ettore accepted the invitation, and at the beginning of March 1936 the final draft of the manuscript was in uncle Giuseppe's hands. Unfortunately, the volume in honour of the famous scholar from Catania was never actually published owing to new budgeting restrictions imposed by the fascist regime, and it only appeared in press after Majorana's disappearance. However, Gentile's choice of the Italian journal *Scientia* was perfectly appropriate, because it was here that well-known economists and sociologists, such as Vilfredo Pareto, Emile Durkheim, and Ernst Cassirer published some of their works.

While the origins of paper P10 very clearly account for the rather different nature of the topics it covers, its content is quite another matter, and more than half a century went by before it was understood and accepted.

To get some idea of what it is about, we may refer directly to Majorana by looking at what he wrote on 27 December 1937 to another uncle of his, Dante Majorana:

> The mathematical method cannot be of any substantial utility to sciences that now have nothing to do with physics. In other words, if one day we discover the mathematical description of the simplest facts regarding our life or consciousness, this will most certainly not happen for a natural evolution of biology or psychology, but only because some further radical renewal of the general principles of physics will extend its domain to fields that still lie outside it. The most relevant example is offered by chemistry which, after remaining an independent science for such a long time and with great glory, has in recent years been fully taken over by physics. This was made possible by the appearance of quantum mechanics, while no useful relationship had been established between chemistry and classical mechanics. While waiting for new miracles to be performed by physics, we should then recommend to scholars of other disciplines to rely on methods proper to their own discipline, and not to look for models or suggestions coming from present day physics, and even less from those coming from previous paradigms. And this, because we still have no feeling at all about the physics that, one day, may have the last word on biological or moral facts (Roncoroni 2012).

This very idea was applied in paper P10. The starting point of Majorana's analysis was a critique of the common idea, prevailing in the 1930s, that the interaction between social sciences and the sciences of Nature should develop along the lines of the celestial mechanics paradigm (whose achievements stretched back over at least two centuries), characterised by determinism, which later became typical of the mechanistic conception of Nature. Indeed, Majorana pointed out that such a deterministic scheme is "subject to a real limitation of principle when we take into account the fact that the *usual* methods of observation are not able to provide us with the exact instantaneous conditions of the system". In other words, Majorana understood that, in the field of social sciences, rather than just taking the analogy with planetary motions, it would be necessary to exploit the complete similarity with what takes place in the statistical mechanical description of a gas (made up of many molecules).

> One should realize that the formal analogy could not be more stringent. For example, when one states the statistical law: "In a modern European society the annual marriage rate is about 8 for 1000 inhabitants". It is clear enough that the investigated system is defined only with respect to certain global characters by deliberately renouncing the investigation of additional information, such as, for example, the biography of all individuals composing the society under investigation. [...] This is not different from when one defines the state of a gas by simply using pressure and volume and by deliberately renouncing investigation of the initial conditions for all single molecules.

What may look obvious today clashed head on with the general vision of mathematical economists in the 1930s, such as Léon Walras, Vilfredo Pareto, Karl Schlesinger, and Abraham Wald, imbued with a strict determinism, which led to the development of the general equilibrium theory (Mantegna 2006).

The general recognition of the role played by statistical laws in social sciences was not, however, the final result of Majorana's analysis, which went far beyond the *classical* vision of such statistical laws. In the 1930s, as a matter of fact, the new paradigm for quantum mechanics had already been developed, and it was right here that Majorana's new conception came into being:

> The statistical laws concerning complex systems known in classical mechanics retain their validity according to quantum mechanics. [...] However, the introduction in physics of a new kind of statistical law or, better, simply a probabilistic law, which is hidden under the customary statistical laws, forces us to reconsider the basis of the analogy with the above-established statistical social laws.

It is indeed the *intrinsically* statistical nature of the examined phenomena—physical or social—that tells Majorana that the statistical laws should be blended into the mathematical modelling of social and economic phenomena, and so acquire the same epistemological status as the irreducible probabilistic laws in quantum mechanics. It is interesting that paper P10 may be considered (Latora 2005) as the first paper about what we now call *complex systems*, and this term also appears explicitly (see above) in Majorana's paper.

In finance, it is easy to understand that such a programme, basing its approach to society on the framework of statistical physics, was only recently transformed from a declaration of principles into a concrete study, the most remarkable result being the Black and Scholes model of option pricing (Black and Scholes 1973), which earned the two economists the Nobel prize. But once again, the story is even more interesting than that.

Paper P10 was completely passed by (if we measure this in terms of the number of citations) from the time it was published until 1997, when a seminal article appeared with the title "Physics investigation of financial markets" (Mantegna and Stanley 1997), opening precisely with a quotation from Majorana's paper. It contained the proceedings of a conference held in Varenna, on Lake Como, by two of the founding fathers of *econophysics* (a term coined by H. Eugene Stanley two years before), and it displayed exactly the same basic programme spelt out by Majorana more than 60 years earlier. However, Majorana's name was associated with these studies only two years later, when a bestseller was published by the same authors, explaining the roots of econophysics (Mantegna and Stanley 1999). This book and other works by the same authors have since received thousands of citations, so it is no surprise that Majorana's paper P10 has rapidly achieved an astonishing popularity among students of complex systems. And of course this is also due to the availability—clearly lacking in Majorana's day—of large databases and the appearance of novel social phenomena based upon the worldwide web, which have allowed scientists to formulate statistical models that can be analysed quantitatively.

It is intriguing that, after the sensational success of econophysics, we find some re-interpretations of the role played by Majorana, linking his name to those of the founding fathers of statistical mechanics, such as James Clerk Maxwell and Ludwig Boltzmann (Castellano et al. 2009). Even more curiously, paper P10 has also stimulated epistemological and philosophical reflection in recent years, for instance, associating Majorana's name with the "objectivist interpretation of social indeterminism" (Boniolo 1987) or again, a "transversal epistemology" (Bontems 2013).

But we shall stop here, for the time has come to return to our story.

Part II
Power to the Italian School

Chapter 4
The 1937 Opportunity

The man of science tries to investigate Nature's mysteries and, when he sees even a pale light, he shares it right away with his peers. That was not the case with the researcher Majorana, as we have already seen. But perhaps something was about to change.

Majorana Applies for a Chair

As already mentioned, the first selection in Italy for a tenure in theoretical physics was held in 1926, essentially through the efforts of O.M. Corbino, and it resulted with the appointment of Enrico Fermi in Rome. The subsequent selection was announced after more than ten years, at the beginning of 1937, this time at the request of Emilio Segrè, who had had the tenure of experimental physics at the University of Palermo since 1936.

In Italy such a long time span was extraordinary for physics (even for theoretical physics), and over this period some very capable young theorists were emerging on the national scene. All of these were benefitting, although often for short periods, from Fermi's guidance in Rome. Their natural ambition was thus to ensure their position in the academic world, so that they could easily and profitably continue their research. Hence, when the 1937 selection came up, it was clear that they would all take part. And this included Majorana, perhaps invited by Fermi and his friends. However, in the light of what was said in Chap. 2, we cannot attach much importance to this sort of "pressure" on Ettore, who seemed, as we have seen, to be very keen on teaching [a fact attested by Amaldi (1968), for example, and Ettore's sister, Maria (Russo 1997)]. Instead the role of Fermi and his friends might have been simply to "guide" Majorana through the selection process, since he was clearly unfamiliar with this kind of thing (for instance, he was awarded the C.N.R. scholarship for his stay in Leipzig only because Fermi took care of it, while Majorana limited himself to writing a short résumé). As a matter of fact, Ettore "had

© Springer International Publishing AG 2017
S. Esposito, *Ettore Majorana*, Springer Biographies,
DOI 10.1007/978-3-319-54319-2_4

not been publishing physics papers for some time. Fermi and his various friends did their best in this direction and Majorana eventually and by a great effort came round to entering the selection; so he published the paper on the symmetric theory of the electron and the positron in *Nuovo Cimento*" (Amaldi 1968). The content of this article, as discussed in the previous chapter, had certainly long been developed, so the "great effort" just mentioned may refer simply to the writing of the paper, since we know Majorana was often reluctant to publish the results of his work, a reluctance he may have overcome this time due to his possible attraction to teaching. In any case, it was Majorana himself who informed Fermi and his friends that he had a paper "ready".

The deadline for the selection was 15 June 1937. For those days, there were certainly quite a few applicants lined up with the required documents[1]: Giovannino Gentile, Giulio Racah, Gian Carlo Wick, Leo Pincherle, and Gleb Wataghin. And, of course, Majorana.

Young Theoretical Physicists

The high professional profile of the applicants for the chair in theoretical physics is suitably expressed in the words of the selection committee, chaired, as we will see later, by Fermi, who knew each of the candidates well. Here are the committee's assessments[2] of Ettore's opponents.

Giovanni Gentile – Graduated in Pisa in 1927; assistant for six months at the Institute of Physics in Rome; winner of a state scholarship, he first attended the institute in Berlin and then the one in Leipzig. He became a lecturer of theoretical physics in 1931 and was appointed for the same discipline from 1931 to 1937, first at the University in Pisa, then in Milan.

He is presenting 12 publications, three of which in collaboration.

In a group of papers relating to spectroscopy and atomic physics, he makes complex calculations to evaluate the energy levels and spectroscopy intensities, reaching interesting and noteworthy results. Another group of papers deals with the difficult problem of ferromagnetic properties; in particular, they examine the relations between the direction of the total magnetic moment and the principal crystal axes, and also the remanence phenomenon. These papers also bring a real contribution to the question of ferromagnetism and reveal a clear physical vision of this complex phenomenon. A huge monograph deals with the diffraction by a thin slit, where the author reaches new results thanks to an exhaustive dissertation, even though the topic is part of classical physics and has already been investigated more than once. Other papers are didactic, educational, speculative, or

---

[1]Besides the application for the selection and the publications, the candidates had to attach a certificate of membership to the Fascist Party (*Partito Nazionale Fascista*), a birth certificate, a general declaration of criminal record, a certificate of Italian citizenship, and a certificate of good moral conduct. For Ettore, these were released respectively on 21, 24, 25 May and 3, 9 June 1937.

[2]See the *Bollettino del Ministero dell'Educazione Nazionale* (Bulletin of the Ministry of Education) (Part II: Administration Records) year 65 (1938), vol. I, p. 280.

philosophical. The lecture courses and monographs testify that the candidate is also a clear and accomplished writer.

In conclusion the Committee is unanimous in recognizing that the candidate has a sound mathematical background, a profound culture in physical matters, and a praiseworthy commitment and physical sense in the study of the problems treated, some of which are among the most difficult in modern physics.

Based on this judgement, the Committee unanimously declares the candidate suitable for the present selection process.

Leo Pincherle – Graduated in physics in Bologna in 1931; winner of a scholarship of the Marco Besso Foundation, he attended the Physical Institute of the R. University of Rome for three years. He became a lecturer of theoretical physics at the University of Padua. In 1936-7 he also taught mathematical physics in the same university.

He is presenting 12 publications. One of these is a course of lectures on experimental physics (mechanics and thermodynamics) given with conceptual precision and clarity; another is the complete description of the means available to produce artificial disintegrations. The other papers present original research. The topics treated deal with problems of spectroscopy, with particular reference to X-rays. The candidate studies several questions regarding the intensities of the lines in this spectroscopic region and, based on a sound knowledge of the quantum theory, is able to bring a remarkable and original contribution to all the problems he deals with, such as the evaluation of the width of the X-ray lines, the intensity of these lines, and the origin of the satellite lines.

The Committee is unanimous in recognizing that the candidate has a sound knowledge and an outstanding aptitude for theoretical research; it must be noted though that his activity has so far been limited to a relatively narrow field.

Based on this judgement, the Committee unanimously declares the candidate suitable for the present selection process.

Giulio Racah – Graduated in Physics in Florence in 1930; he attended the Physical Institute of the R. University of Rome in 1930-1; in 1931-2 he won a state scholarship and went to Zurich. Lecturer in theoretical physics since 1933, he has been in charge of the course in theoretical physics at the R. University of Florence since 1932, and the same course at the University of Pisa in 1936-7.

He is presenting 24 theoretical publications, one in collaboration. Some of these papers are mainly mathematical and attest to a high level of knowledge in this field, and particularly to the candidate's mastery of the difficult methods of group theory. Two papers bring a notable contribution to the proof of the equivalence of Dirac's radiation theory with the classical theory when considering interference phenomena. The candidate has investigated the problem of hyperfine structures, bringing an important contribution to the theoretical interpretation of these phenomena. Lastly, a considerable number of papers use a range of mathematical methods to study the theory, of fundamental importance, of the scattering of fast particles. From the analysis of these papers, which deal on the whole with problems of real physical interest, it is clear that the candidate has a vast physical knowledge and also the necessary analytical ability to perform particularly difficult calculations. In a conference on the nuclear theory, he has shown himself to be a gifted speaker, with uncommon didactic abilities.

The Committee is unanimous in recognizing that the candidate stands out among Italian experts of theoretical physics.

Based on this judgement the Committee unanimously declares the candidate suitable for the present selection process.

Gleb Wataghin – Graduated in physics in Turin in 1922; he then graduated in mathematics in 1924. He was assistant to various professors, between 1922 and 1928, at the R. Higher Institute of Engineering in Turin, and in 1930 he became a lecturer in theoretical physics. He has been appointed to different positions. At present he is professor of experimental physics at the University of São Paulo in Brazil. In 1929, he was awarded the first prize in a competition run by the Accademia Pontificia dei Nuovi Lincei.

He is presenting 57 publications, of which some are in collaboration. Most of these publications are theoretical, some experimental, others didactic. Among the former the Committee notes the following: the correct treatment of the statistics of an electron gas in the presence of pair formation; a large set of papers attempting to eliminate the known difficulties with the relativistic treatment of quantum electrodynamics, using a procedure which nevertheless does not yet seem fully justified. Other papers deal with the uncertainty relations, the relativistic equations of a massive particle, and related topics. Although these papers prove the candidate's vast knowledge of the hardest topics of theoretical physics, and a commendable enthusiasm for research, the Committee cannot see any essential contribution to the solution of the treated problems; on the other hand, we should bear in mind that these problems involve very serious difficulties, as everybody knows. The experimental papers deal with an accurate study of the reflection of light emitted by a moving source, two methods for producing light modulation, and a study of reflection from a vibrating piezoelectric quartz. Apart from his long teaching experience, Wataghin's sound presentation skills are attested by his critical dissertation on quantum theory, rewarded by the Accademia Pontificia dei Nuovi Lincei, as well as in the courses he gave in Turin and São Paulo.

For the candidate's overall scientific and didactic activities, the Committee unanimously declares him suitable for the present selection process.

Giancarlo Wick – Graduated in physics in Turin in 1930; winner of the Ugo Fano scholarship, he attended the universities of Göttingen and Leipzig during one semester. He has been assistant at the Institute of Physics in Rome since December 1931. Lecturer in theoretical physics since 1935, responsible for physics at the Higher School of Architecture in Rome since 1935.

He is presenting 20 publications, of which three in collaboration.

Two of these are experimental and contain very important results on the scattering of different groups of neutrons. Another paper deals theoretically with the action of an electric field on the higher order terms of spectral series, interpreting the experimental results in a satisfactory manner. The topics of the other papers concern the most important questions in modern theoretical physics. In all the papers the candidate shows a sound knowledge of analytical methods, developed with exceptional elegance, and also the genuine physical aspects of the questions; for each problem treated he has made very valuable contributions. Particular note should be made of the following: the calculation of the lifetime of a radioactive element due to positron emission, the study of the scattering of slow neutrons, the study of the magnetic moment of the hydrogen molecule—where he brilliantly explains a discrepancy between the experimental and theoretical values—, and the investigation of the magnetic moment of the proton as inferred from the theory of beta decay. In his expository papers he shows outstanding didactic skills.

All these results show that this candidate holds a remarkable position among Italian experts of theoretical physics, and the Committee unanimously declares him suitable for the present selection process.

Considering these appraisals, together with what has already been said about Majorana, it is very clear what a difficult job it would be to choose only one candidate for the chair of theoretical physics (or even to choose a shortlist of three winners) among these gifted participants. However, surprises were just around the corner.

High and Well-Deserved Repute

Before the summer break, the Minister of Education Giuseppe Bottai appointed the tenure selection committee according to the laws in force, and in August Majorana and the other applicants already knew its members. They were Antonio Carrelli and Enrico Fermi (secretary and president, respectively), Orazio Lazzarino, Enrico Persico, and Giovanni Polvani. The committee met for the first time on Monday, October 25 at the Institute of Physics in Rome, and here is what we find in the minutes:

> After a thorough exchange of ideas, the Committee is unanimous in recognizing, among the other applicants, professor Majorana's absolutely exceptional scientific standing. The Committee has thus decided to send a letter and a report to H.E. the Minister to propose Majorana's appointment as professor of theoretical physics for his high and well-deserved repute in one of the kingdom's universities, independently of the selection process requested by the University of Palermo.[3]

The initiative suggested by the committee referred to the special merits mentioned in clause 8 of the *Regio Decreto Legge* (Legislative Decree) of 20 June 1935, no. 1071, introduced some years earlier to allow the Nobel prize-winner (1909) Guglielmo Marconi (who did not have any academic position) to obtain, without the usual selection, the chair of electromagnetic waves at the University of Rome. Such a decision was clearly well thought out if, as written in the minutes, the "thorough exchange of ideas" lasted for three whole hours (from 4 to 7 pm). Part of this time, however, was spent writing the letter to the Minister, and above all the report about Majorana's scientific activity. Without drawing attention to the scarceness of his publications, they needed to stress the importance of those papers he had actually published.

The justification for such an unusual procedure, undoubtedly motivated in itself but unnecessary here (Ettore would have been the winner by a wide margin), might

---

[3]See the minutes of the Committee for 25 October 1937 quoted in Recami (1987).

be as explained by G.C. Wick more than 50 years later.[4] If Majorana had been first
on the list, the final shortlist of three would not have included Giovannino Gentile,
the son of the well-known philosopher and senator, former Minister of Education.
Given his political importance, he might have put "pressure" on the committee,
suggesting a possible alternative with Majorana's nomination "*hors concours*".
Such a motivation is highly plausible, but unfortunately not well documented. On
the other hand, there is strong evidence to support this hypothesis.

To begin with, Giovannino Gentile was certainly worried about being "left out"
of the selection; he spoke about it openly with his friend Ettore, who gave the
following affectionate answer in a letter on August 25:

> I believe your deliberate suspicion towards Fermi is unnecessary; he spoke to me about you
> with the most straightforward fondness. As to the other members of the committee, either I
> have never seen them or have not seen them for ages, but it seems to me that at least one of
> them should have the authority and the will and the duty to stand by Giovanni Gentile.[5]

Giovannino's apprehension may have been revealed to his father as well or,
more likely, the senator himself may have done his best to collect "information" on
his son's chances of winning. Now, on the one hand it is reasonable to think that the
"authorised personnel"—those who had the task of evaluating—already had quite
clear ideas regarding the skills and priorities of the different applicants. On the other
hand, it looks odd that the story of the third person being excluded from the
winning three, as narrated by Wick, was known to Gentile Sr., too. Unless the
senator had not been informed by one of the "authorised personnel". Among these,
Fermi for sure—and consequently also Persico, who was very close to him—knew
all the candidates sufficiently well. This probably applies equally to Carrelli who, as
we will later see, though based in Naples, had long frequented the Rome
group. Wick's account of the selection process ends symbolically, recalling that
"Fermi did not like to be put under this or any other kind of pressure regarding the

---

[4]See Russo (1997). Segrè (1993) also writes:

> Initially, I had expected that the three winners would be Wick, Racah, and Giovanni Gentile
> Jr. I never dreamed Majorana would enter the competition, because he had lived in
> seclusion for several years. Completely unexpectedly, however, he did. The consequence
> was clear: the three winners would be Majorana, Wick, and Racah; Gentile would be left
> out. In a theoretical physics selection committee, the opinions of Fermi and Enrico Persico
> would be decisive, and both would honestly recognize merit. [...] I believe, on good
> grounds, that in order to avoid a defeat for his son, Gentile's father, a former minister of
> education and still a power in Italian politics, had conceived this plan and suggested it to the
> committee. With the selection held in abeyance, Majorana was appointed professor at
> Naples based on exceptional merit. A law allowed for this procedure in special cases
> involving illustrious persons, and had been used, for example, in the case of Marconi. After
> Majorana's appointment, the selection process was reinstated, obviously without Majorana's
> candidacy. The three chosen were Wick, Racah, and Gentile. To my delight, Wick came to
> Palermo not long thereafter. Needless to say, at the time I was completely in the dark about
> the maneuvers mentioned above.

[5]Letter MG/R6 of 25 August 1937 in Recami (1987).

results of the selection" (Russo 1997), but we cannot help noticing, in retrospect, that the Institute of Physics in Naples directed by Carrelli (who was appointed secretary of the selection committee) greatly benefitted from the situation,[6] as we will see shortly.

The letter and the report written by the committee on October 25 were sent immediately to the Minister Bottai who, accepting the proposal, sealed the letter in his own hand with the word "Urgent" (Recami 1987). And this was the way it was in fact dealt with.

On November 2 the Minister issued the decree to appoint Majorana "independently of the normal selection procedure, as professor of theoretical physics at the Faculty of Science of the Royal University of Naples [...] effective from November 16, 1937". At the same time, he informed the chancellor Giunio Salvi of the Neapolitan athenaeum of his decision; Salvi immediately sent a letter of invitation to the new professor, who answered only two months later, however, when he arrived in Naples and found the letter there (Recami 1987).

On December 4 the Minister's decree was registered at the *Corte dei Conti* (National Audit Office) and it became effective; the appointment was later announced to Majorana at his home in Rome on 10 January 1938.

However, Ettore and the other applicants were informed indirectly well before that (in fact, at the beginning of November) and Ettore himself was so surprised that he wrote to his uncle Quirino on November 16:

> I had a bit of a laugh about the oddities concerning my selection, which I was totally unaware of. I do hope I will go to Naples.[7]

So Ettore was genuinely pleased with the prospect, which once again confirms his ambition to engage upon a teaching career, and Carrelli was probably contributing to that. In another letter a few days later, on November 21, leaving out the ritual scepticism in favour of his subtle irony, he wrote to his friend Gentile:

> I still do not know whether and when I will go to Naples. I am in contact via letters with Carrelli, who really is a decent person (his motto: men are much better than we think). Segrè and all the others have also been very kind. I am surprised that as far as I am concerned you doubt my guts, in the metaphorical sense. [Pope] Pius XI is very old and I have received a very good Christian upbringing; if in the next conclave I am appointed Pope for exceptional merit, I am definitely going to accept. Excuse me now if I stop at the first half kilometre. My affectionate greetings and best wishes to you.[8]

The "wishes" were probably propitiatory for Giovannino, for whom the "ordinary" selection process was not yet finished.

Once the "Majorana affair" had been solved, the prestigious committee was able to resume its task, and after writing the appraisals quoted above, decide the final

---

[6]Though in a different context, one of Carrelli's (and Majorana's) students has suggested that Carrelli and the Senator Gentile were more than just simple acquaintances. However, we have no evidence that this happened before the events discussed here.

[7]Letter MQ/R29 of 16 November 1937 in Recami (1987).

[8]Letter MG/R7 of 21 November 1937 in Recami (1987).

three winners: G.C. Wick, G. Racah, and G. Gentile, in order of merit. As requested, the first went to teach at the University of Palermo, while the other two were later given posts in Pisa and Milan, respectively.

Ettore Majorana thus became a professor at the University of Naples, "for the high repute of the particular expertise he has gained in the field of theoretical physics".[9]

Great Expectations

The main virtue of the "wheelings and dealings" of the 1937 selection process was that, besides giving Majorana his rightful and well-deserved recognition, a long-standing post of theoretical physics was finally set up at the University of Naples. This discipline had already been part of the curriculum at the Faculty of Science but, like the majority of physics subjects (taught not only in this faculty, but also in the engineering and medical faculties), it was taken care of by the only full professor, who was also director of the Institute of Experimental Physics, namely, Carrelli, while the other physical disciplines depended on the director of the Institute of Earth Physics, who was Giuseppe Imbò at that time. As we will see in more detail later on, Carrelli was one of the "new generation" of physicists, not just because he was only a year older than Fermi (and six years older than Majorana), but because his scientific interests turned mainly around the new quantum physics which was catching on at the time. In particular, his papers on atomic and molecular spectroscopy were appreciated even by Fermi, and he cited them in some of his papers. But, according to his former students, his course on theoretical physics was not "modern" enough, and certainly not as cutting edge (in Italy) as Fermi's in Rome. Conceivably, the reason was that, though he had often worked on theoretical problems throughout his long scientific life, Antonio Carrelli was essentially an experimental physicist. Therefore, even when faced with a new teaching commitment, this would focus on experimental rather than theoretical aspects. Majorana himself, in a letter to his family dated 11 January 1938, noted that "Carrelli prepares his lessons on mechanics with many little games",[10] that is, with the help of experimental tools, and that "the prevailing activity is practice".

He was probably aware of his limitations and may have been hoping that the same thing would happen in the Institute of Physics in Naples as had happened ten years before when Corbino had asked Fermi to go to Rome. That is why, with the help of Fermi himself, Carrelli quickly took (if not actually created) the opportunity of having a "genius" like Majorana in his institute. And even on 21 November 1937, Majorana was already writing that he was "in contact via letters with Carrelli". This was a relationship, not only via letters, which was clearly established

---

[9]See the Minister's announcement letter D/ME4 in Recami (1987).
[10]Letter MF/N1 of 11 January 1938 in Recami (1987).

by the man from Naples, given Ettore's extremely reserved and introverted character. Anyway, probably helped along by the proverbial Neapolitan joviality, Majorana immediately liked Carrelli, and we may say that this was attested on both sides. "Carrelli is really a decent person", Ettore wrote in the letter to his friend Gentile, while Amaldi, who did not know about this letter yet but had close relations with Carrelli (after the war), also says that "in Naples [Majorana] became friends with Antonio Carrelli" (Amaldi 1968).

Favourable conditions were developing, unexpectedly and quickly, for a significant revival of physics in Naples. And Majorana, at least at the beginning, did not shrink from these expectations: "I will give my whole energy to the school and to Italian science, which is fortunate enough today to be rising towards its former leadership".[11]

---

[11]See the thank you letter MM/N to the Minister of Education of January 12, 1938 in Recami (1987).

Chapter 5
Arrival in Naples

The First Three Days

The academic year of 1937–8 began at the University of Naples on Friday, October 29, while the Minister Bottai was evaluating the "particular expertise" of the candidate Majorana. So some courses had already been running at the Institute of Physics for some time when they resumed on Monday, 10 January 1938, after the Christmas break. The academic holidays actually lasted from Wednesday, December 22, to the following Saturday, January 8, which was "Her Majesty the Queen and Empress's birthday". So it was then that Ettore Majorana arrived in Naples, a new professor of theoretical physics, moving into lodgings at the *Albergo Napoli*.

He took his first steps in the premises in *Via Tari* under the supervision of the director Antonio Carrelli, who had already been in contact with Majorana and took him right away to "report for work" to the Dean of the Faculty of Science, Umberto Pierantoni. Together they arranged that the course should start on Thursday, January 13. Pierantoni also did everything he could to help the exceptional new arrival to settle in at the Institute. "Carrelli has been very kind, and today we bought some furniture for my room, kindly offered by the Faculty",[1] Majorana wrote to his mother on Tuesday, January 11.

Then Carrelli showed him around his institute and introduced the few collaborators who worked there. "The Institute is very clean and tidy, though poorly equipped", we also read in the above-mentioned letter; there is "actually only Carrelli, his old chief assistant Maione, and the young assistant Cennamo". The lack of means and people clearly struck Majorana, who had for so long frequented Corbino's well-endowed Physical Institute in the *Via Panisperna* in Rome. But this might well be considered of less importance by a theoretical physicist who was used to working alone and who had come to Naples to teach. He was probably more

---

[1]Letter MF/N1, *loc. cit.*

© Springer International Publishing AG 2017
S. Esposito, *Ettore Majorana*, Springer Biographies,
DOI 10.1007/978-3-319-54319-2_5

interested in his students, and yet he would have to wait a little longer to make their acquaintance; for the time being he had to make do with what information Carrelli could give him about them.

The theoretical physics course began on Thursday, January 13, at 9 o'clock. In 1938 the famous Imbriani Act of 16 February 1861 was still in effect, although not for much longer: it gave greater freedom to the University of Naples than to other Italian universities, and allowing it to maintain far more ancient customs (since the Bourbons or even earlier). One of these was the *Lectio magistralis* a new professor had to give in front of other Neapolitan scholars, whose duty was to evaluate his suitability. Majorana's first lesson at the Neapolitan University should have served this purpose, but in a letter to his family on January 11, Majorana himself tells us that he has arranged with the Dean of the Faculty "to avoid every official character at the opening of the course".[2] He nevertheless prepared, perhaps the day before the lesson, some personal notes for his own inaugural lecture, in which he gave advance notice of his objectives and the methods he would employ in *his* course in theoretical physics, even using a fairly dignified tone here and there:

> In this first introductory lecture I will briefly discuss the aims of modern physics and the significance of its methods, with particular emphasis on their most unexpected and original aspects with respect to classical physics. Atomic physics, which will be the main subject of my discussion, despite its important and numerous practical applications – together with those of a wider and perhaps revolutionary impact that the future may have in store –, is first of all a science of immense speculative interest for the depth of its investigations, which truly reach down to the ultimate roots of the facts of nature (Esposito 2006b).

The introductory lecture was held in the great hall of the institute in *Via Tari*, but it was probably attended by little more than ten professors, and perhaps none of the students, at least not the "official" students who would attend the course. Majorana was puzzled by such a "disappointing" welcome, as some of his former students recall, because he had not seen that coming, although in the letter to his family two days previously, we can already find a possible explanation:

> It has not been possible to check out whether there are overlapping lessons, so the students may not be able to come and we may be forced to postpone.

But in the end, nothing was postponed and everything happened as we have just described. This included Ettore's mother and other relatives coming down to Naples for the introductory lecture, as his sister Maria remembers, a journey Majorana had particularly advised against. Without success.

---

[2]Things were different for G.C. Wick, who was granted the tenure of Theoretical Physics in Palermo. In *Il Giornale di Sicilia* of January 23, 1938 we find:

> Royal University of Studies – Tuesday 25, 1938, at 11 o'clock, in the Graduation Hall the Illustrious Professor Gian Carlo Wick will deliver the introductory lecture to his official course on "Some aspects of modern physics".

Innocent Souls

The course began the following Saturday, January 15, and it was attended only by those directly interested in Majorana's lessons. For the occasion they all had to move to a smaller classroom opposite the great hall used two days earlier, on the ground floor of the institute. This classroom gave onto the inner courtyard of the university and was reached along a corridor where the light came in through large windows. There were few students, but Ettore might well have expected this, since not so many people were attracted by a career as a "physicist". The official numbers were five students, including four women and a man: Filomena ("Nella") Altieri, Laura Mercogliano, Nada Minghetti, Gilda Senatore, and Sebastiano Sciuti. One should not be surprised by the greater number of female students, although it was strange at the time. As a matter of fact, it is of no statistical significance: in the academic year 1934–5, when Majorana's students enrolled, the degree course in physics in Naples saw 15 men and 4 women register, and the proportion remained more or less the same in the following years. If anything, what is worth noticing is that *all* the female students registered for that year would appear to have attended Majorana's course.

The students came from a variety of backgrounds. Minghetti, for instance, was born in 1917 near Ancona; her family was from Turin and she studied in Naples. Sciuti, on the other hand, was born in Naples, also in 1917, and in his own words[3] he enrolled in physics just because of the brilliant prospects for Italian science under Fermi's guidance, something Corbino had already emphasised more than ten years before. So he was particularly keen on attending an "advanced" course given by a member of the celebrated Rome group. Gilda Senatore, perhaps the only one really gifted in theoretical physics, was born in Sao Paulo, Brazil, in 1914, but her family was from a small town near Cava de' Tirreni, in the province of Salerno, to which they returned when she was 8–9 years old. Her father Geremia was a chemical engineer (one of the first in Naples), but did not approve of his daughter studying at university. However, her uncle Giovanni Pisapia, one of her late mother's relatives, helped her to go to university, and paid the costs of the fees and the daily journeys from Cava de' Tirreni to Naples. According to Senatore herself,[4] her uncle Giovanni used to meet professor Carrelli at the theatre, and this might have been a reason why his niece enrolled in physics.

Ettore thus entered this environment, which was not obviously reaching towards any "depth of scientific enquiry" of the kind he had mentioned in his inaugural lecture. And the knowledge the students had acquired in their previous years of study was not promising at all. In January 1938 they were attending their fourth year, the last one in their course of studies, the first three being organised as follows:

---

[3]C. Bernardini and L. Bonolis's interview with S. Sciuti, July 2001. We thank L. Bonolis who kindly made the transcript of the interview available.

[4]S. Esposito's interview with G. Senatore, February 2004.

1st year:

- Experimental physics
- General and inorganic chemistry
- Algebraic analysis
- Analytic geometry
- Projective geometry
- Practical preparation of chemistry experiments.

2nd year:

- Experimental physics
- Organic chemistry
- Mineralogy
- Infinitesimal calculus
- Practical analytical chemistry
- Practical physics.

3rd year:

- Rational mechanics
- Theoretical physics
- Practical and research physics
- Two other subjects chosen among: technical physics, physical chemistry, geodesy, electrodynamics, advanced calculus, further mathematics.

Although some of these courses were given by important scientific personalities, such as Francesco Giordani for chemistry, Gabriele Mammana for algebraic analysis, Antonio Signorini for rational mechanics, and Renato Caccioppoli for projective geometry, the preparation provided by the five "innocent souls", as one of them[5] would so charmingly refer to them, was not at all appropriate for Majorana's course. The students themselves, after attending Carrelli's two-year course in theoretical physics the previous year, felt that the mathematical tools at their disposal were totally inadequate to ascend the peaks—that is how they saw them—that their teacher, or more accurately, modern physics, was leading them to.

Majorana was fully aware of this, as he had already realised in Rome that modern quantum mechanics required a depth of analysis that was not yet the norm. But it was a depth he had every intention of reaching in his course on theoretical physics, as so clearly stated in the plan he declaimed on January 13. The appropriate mathematical aids were introduced by Majorana himself (in a straightforward but, at the same time, accurate way) in the second part of the course. But not before thoroughly discussing all those phenomenological aspects—typically physical rather than mathematical—that would underpin the later developments of this new physical enquiry, presented in a way that clearly echoed Fermi's course,[6] the one

---

[5]S. Esposito's interview with G. Senatore, *loc. cit.*
[6]The broad lines can be found in the textbook prepared by Fermi himself (Fermi 1928).

Majorana had attended as a student.[7] The constant switching from physics-related topics to more formal ones was a permanent feature of Majorana's lectures in Naples, paying particular attention to didactic presentation and comprehension.

Nevertheless, the students who had that lucky opportunity would meet with quite serious difficulties, perhaps also because of worries over the exams they had put off. Majorana noticed this and confessed his concern to Carrelli (Amaldi 1968). However, as the course went on, the situation probably improved, thanks also to the notes provided by their teacher; in a letter to his friend Giovannino Gentile written on March 2, Ettore wrote that he was "happy with the students, some of whom are going to take physics seriously".[8] In retrospect, we know that all five of the students graduated in December 1938 (except for Senatore, who graduated a year later), and after a short period following graduation, never had another opportunity to do physics research, with the exception of Sciuti, who first became an assistant, after moving to Rome, and then professor at the university there. Quite plausibly, in the letter to Gentile, Majorana was in fact referring to Sciuti; as recently documented (Esposito 2005b; Drago and Esposito 2007), Sciuti was the only one who dared ask the professor for explanations, and he was addressed in a "particular" way in the letter to Carrelli of March 25.[9] Apart perhaps from the beautiful student Gilda Senatore, to whom Majorana gave all his papers before disappearing.

A Mathematician's Interest

The audience for Majorana's lectures on theoretical physics in the little classroom was slightly larger than the "official" group of five students, because most of the time they were joined by three guests who were more or less curious about the new course. While Mario Cutolo was probably there because he was more interested in a girl (N. Minghetti) than in Majorana's lessons, the other two, Father Savino Coronato and Eugenio Moreno, were actual "emissaries" of the outstanding mathematician Renato Caccioppoli. The latter knew Carrelli very well, not only for academic reasons, but also because they both attended Pietro La Via's cultural salon in Posillipo (Toma 2004), and he very likely got the news of Majorana's appointment as professor of theoretical physics for exceptional merit from Carrelli himself.

---

[7]See (De Gregorio and Esposito 2007). There is much common ground between the first part of Majorana's course in Naples and Fermi's in Rome, where the quoted book was in use. Clearly, the phenomenological treatment of the old quantum theory, as developed by Fermi, was considered adequate by Majorana, and he employed it in his course in Naples, too. However, compared to Fermi's course, Majorana intended to go well beyond, both as regards the mathematical formalism chosen and the topics included, but he only accomplished part of his project, due to the early interruption caused by his disappearance.

[8]Letter MG/N1 of 23 March 1938 in Recami (1987).

[9]Letter MC/N of 25 March 1938 in Recami (1987).

The mathematician's interest very likely sprang from the fact that he had heard the introductory lecture of January 13, which he would probably have attended as a professor of the Faculty of Science. It is also curious to note that Caccioppoli had taken part in the debate some years earlier among Italian mathematicians regarding the solution of the Thomas-Fermi equation (Di Grezia and Esposito 2004), as already mentioned above. Caccioppoli's participation was not direct, but came rather through the involvement of Carlo Miranda and others. This is curious because he (and the other mathematicians) did not realise that the question had already been dealt with well before and by Majorana himself.

In any case, regardless of the probable interest in the presence in Naples of a theoretical member of Fermi's group, or even of Majorana's introductory lecture, the fact remains that Caccioppoli's friends and colleagues confirm that he knew and regularly met up with Majorana during his stay in Naples (Toma 2004); but we cannot help pointing out the completely different characteristics and personalities of the two great minds. It is also a lucky and interesting fact, as we will see, that two of Caccioppoli's students attended the lessons on theoretical physics. Although in very different ways. One of them, Father Savino Coronato, born near Salerno in 1908, was in the last year of his mathematics studies and would graduate in 1938, later becoming Caccioppoli's trusted assistant; his presence at Majorana's course has been attested by Senatore and Sciuti. The other, the Neapolitan Eugenio Moreno, was two years younger than Coronato and, as he had enrolled in mathematics as far back as 1929–30, went well over the prescribed time for his course (he graduated in mathematics in 1940) because he had meanwhile been involved in military activities. His presence at the course is attested by his recollections to his children and mathematical colleagues, and above all—but not only—by the fact that he kept the transcript of his notes from Majorana's lessons, known today as the Moreno Paper.[10]

75 Days in Naples

The reconstruction of the days Majorana spent in Naples may not just be of scientific interest (traceability of the author's scientific documents still unknown), but also useful to correctly establish the context of his ensuing disappearance. Until recent times, such an analysis has revealed features that are largely incomplete or else based on shaky grounds; but the direct and indirect information we learn from the Moreno Paper or other more or less well known sources allows us now to present some sensible and interesting results relating to the 11 weeks Ettore stayed in Naples.

---

[10]See the introduction in (Esposito 2006b).

A somewhat general consideration, reported by Amaldi and probably borrowed from Carrelli, may be usefully stated here:

> In Naples, as after all in Rome, he lived an extremely secluded life; in the morning, when he had lessons, he would go to the Institute and, in the late afternoon, he would walk in the liveliest areas of the city (Amaldi 1968).

It is most unlikely that Majorana took any part in the city's cultural activities, such as mixing in with society circles like his colleague Caccioppoli and the director Carrelli. It is more likely, instead, that Ettore devoted most of his time to preparing his lessons: "As he had always done for all his duties in the past, he carried out the task of giving lessons with great care and diligence". In particular, after a certain time, he also dealt with writing all his lesson notes.

A DAILY DIARY

January 10, Monday: Majorana receives a letter from the Minister Bottai, stating that he has been officially appointed full professor at the University of Naples.

January 11, Tuesday (or the day before): Majorana arrives in Naples from Rome and gets a room at the *Albergo Napoli*[11] or, more precisely, at the *Hotel de Naples*, in *Corso Umberto I 55*, right opposite the central university building.

He visits the Institute of Physics and meets the Dean of the Faculty of Science, Umberto Pierantoni, to arrange the beginning of the course. At the institute he finds, waiting for him, a letter in which the chancellor Giunio Salvi announces his appointment as full professor; he then replies to this letter. Once back at the hotel, Ettore writes the first letter to his mother from Naples.

January 12, Wednesday: He goes to the institute and writes a letter of acknowledgment to the Minister Bottai (using the institute letterhead).[12] He probably writes the notes for the inaugural lecture.

January 13, Thursday: At nine o'clock he gives the inaugural lecture for the theoretical physics course in the great hall of the Institute of Physics; his family, or at least his mother and his sister Maria, arrive from Rome to attend the lecture.[13] Following that, and perhaps on his mother's advice, Ettore probably leaves the *Albergo Napoli* and moves to the *Hotel Terminus*, next to the Central Station.[14]

January 15, Saturday: In the little theoretical physics classroom he gives the first actual lesson with the students (lesson no. 2). The following lessons are usually given in the same room on Tuesday, Thursday, and Saturday mornings.

January 17, Monday: In front of the chancellor G. Salvi and two witnesses, G. Quagliariello and P. Monomi, he swears loyalty to the King and the Fascist Regime,

---

[11]Letter MF/N1, *loc. cit.*

[12]See the acknowledgement letter MM/N to the Ministry of Education of January 12, 1938 in Recami (1987).

[13]Maria Majorana's statement in Recami (1987).

[14]See the letter to his mother of January 22, which may well be seen as Ettore's first contact with his family after he came to Naples for the introductory lecture, where he hints at the fact that he is *still* at the Hotel Terminus.

as required of all the teachers according to the laws in force (clause 18 of Royal Act of 28 August 1931, no. 1227).

January 18, Tuesday: After lesson no. 3, Ettore meets the experimental physicist Giuseppe Occhialini, a year younger than him. Occhialini remembers that (Russo 1997):

> I met Majorana on my way back from Brazil to Italy, during the university break, in January 1938. The ship's final landing was Trieste, but with a half-day stop in Naples, and so I had decided to go and meet professor Carrelli at the Institute of Physics. As a matter of fact, as soon as I arrived in Naples, I rushed to meet him. When I arrived at the Institute it was almost one o'clock and I saw professor Carrelli about to leave. We had quite a light conversation, and then there came a young man, dark eyes, dark hair, I thought he was a student. We introduced ourselves... it was Majorana.

This evidence, which has come out only recently, is confirmed (and elucidated) in one of the Neapolitan newspapers of the day (*Il Mattino*). Here we read that, on Tuesday January 18, 1938, the ship *Oceania* docked in Naples, from South America. The arrival was due at 7:00 o'clock, and then at 3:00 p.m., the ship continued on its journey to Trieste, with no other stop. Given that docking at an Italian port required some customs operations, Occhialini's reference to a meeting with Carrelli and Majorana at about lunchtime seems reasonable (the port is not far from the institute, in fact, about fifteen minutes' walk).[15]

January 20, Thursday: He gives lesson no. 4 and fills in the employee register as a new teacher (Recami 1987).

January 22, Saturday: After lesson no. 5, he writes (from the institute) the second letter to his mother,[16] in which, among other things, he says he has hired a nurse (perhaps to cure his gastritis), and asks his brother Luciano to withdraw money from his bank account in Rome.

January 23, Sunday (or the following day): A probable trip to Rome to get books from his home and from the bookshop Treves. At the same time, it should be noted that for the lessons following this date (from no. 6 on January 25) Majorana writes the notes for the course students.

January 29, Saturday (or the following Monday, 31): Majorana may have gone to the Bank of Italy (whose Neapolitan branch is on the corner between *Piazza Municipio* and *Via Cervantes*, not far from the institute) to withdraw his salary.[17]

---

[15]Other possible dates for Occhialini's meeting with Majorana could only be Thursday, January 20, and Tuesday, February 22. On the former date he would have arrived (at 7 am) in Naples from North America on the ship *Saturnia*, which would later (at 1 pm) go on to Patras, Raguso, and Trieste. If we stick strictly to Occhialini's statement (and to simple "time" considerations), the date suggested here seems to be the only possibility.

[16]Letter MF/N2 of January 22, 1938 in Recami (1987).

[17]Note that, in that period, salaried workers withdrew their salary on the 30th of the month (or, if it was on a Sunday, the previous day). Saturday was a working day for the Bank of Italy, though the opening times were shorter (only in the morning).

From this date on, and until about Sunday, February 20, Ettore probably goes to stay for the weekend with his family in Rome, as shown indirectly by the fact that there are no letters to the family during that period.

February 1, Tuesday: Fifteenth anniversary of the foundation of the Fascist Militia, whence he does not give a lesson.[18]

February 19, Saturday: Majorana does not give the scheduled lesson (we do not know the reason why, but see below).

February 22, Tuesday: Bruno Mussolini and his "*sorci verdi*" ("green rats") disembark in Naples—the newspapers have photos of the event, showing huge crowds gathered in the *Piazza Municipio*, near the port. The gathering of the *Fasci Universitari* (University Fascist Groups) is announced and Majorana does not give his lesson.

February 23, Wednesday: He moves to the *Albergo Bologna* (or, more precisely, the *Hotel Bologna e Grande Bretagne*, in *Via Depretis 72*), situated between the natural continuation of the *Corso Umberto I* (that is, Via *Depretis*) and a street that crosses it, which Ettore originally overlooked, before changing room. Here he writes the third letter to his mother.[19]

From February 24 to the following March 2 there are no lessons at the university, as it is an academic holiday for the Carnival celebrations.

February 28, Monday: He probably goes to the Bank of Italy to withdraw his salary.

March 1, Tuesday: Shrove Tuesday, a day Majorana spends in Naples (see the letter to Gentile of the following day).

March 2, Wednesday: He most likely goes to the Institute of Physics, where he writes a letter to his friend Giovannino Gentile as a response to a postcard sent from Rome.[20] On this day the poet Gabriele D'Annunzio dies, an event given much coverage in the press.

March 5, Saturday: In contrast to what he said to Gentile, Majorana does not give the scheduled lesson, perhaps due to an official directive linked to D'Annunzio's death.

March 8, Tuesday: He resumes lessons (from no. 16), after the long Carnival break, and begins the second part of the course, introducing the more formal (mathematical) tools appropriate for quantum mechanics.

March 9, Wednesday: He writes the fourth letter to his mother from the hotel.[21]

March 12, Saturday (or the following day): After lesson no. 18, he probably goes to Rome to his family (and this will be the last time he sees them).

---

[18]By the way, it should be noted that from February 10, because of the renovation works for Hitler's forthcoming arrival in Naples (the following May), *Corso Umberto I* was closed to traffic; this was the road where the University central building was located and which probably Majorana usually walked along to get to the institute.

[19]Letter MF/N3 of 23 February 1938, 1938 in Recami (1987).

[20]Letter MG/N1, *loc. cit.*

[21]Letter MF/N4 of 9 March 1938 in Recami (1987).

March 15, Tuesday: King Vittorio Emanuele III visits Naples and the *Maschio Angioino* castle, where he inaugurates the exhibition *Three centuries of Neapolitan paintings* (XVII–XIX centuries). There are no lessons at university.

March 19, Saturday: The university is closed for Saint Joseph's day festivities, and Majorana does not give a lesson. From the hotel, he writes the fifth letter to his family (addressed to his brother Salvatore, and not to his mother) and a telegram in which he says he will not go to Rome for the weekend.[22]

March 21, Monday: He probably carries out "errands at the General Register Office and somewhere else", as hinted in the letter of the previous Saturday.[23]

March 24, Thursday: He gives his last lesson (no. 21) to the students at the Institute of Physics.

March 25, Friday: In the morning he goes to the institute to hand in a file of lecture notes (perhaps also containing other notes of personal study) to the student Gilda Senatore (according to her own recollections). Back at the *Albergo Bologna*, he writes a letter to Carrelli[24] and another to his family[25] (this last one will be found in his hotel room some days later) where he announces his decision to "disappear". In the afternoon, at about 5 pm,[26] he departs from the hotel, leaving there all clothes and papers; he then sends the letter to Carrelli. At 10:30 pm he PROBABLY boards the ship *Città di Palermo*, which carries out the Naples to Palermo postal service on behalf of the shipping company *Tirrenia*.

March 26, Saturday: He PROBABLY arrives in Palermo in the morning, where he gets a room at the *Grand Hotel Sole*, in *Via Vittorio Emanuele 291* (about two kilometres away from the maritime station); from here he writes a telegram and a letter through priority mail to Carrelli, where he announces his intention to go back to Naples, but to give up teaching.[27] Then he cables the *Albergo Bologna* in Naples to keep his room.

According to professor Vittorio Strazzeri[28] (professor of geometry at the University of Palermo), in a letter to Ettore's brother Salvatore Majorana dated May 31, Ettore actually boarded the ship to go back to Naples, because he travelled with him in the same three-bed cabin (the third bed was taken by a certain Carlo Price, a foreigner but with a southern Italian accent).

Ettore PROBABLY arrives in Naples on the following morning; and this is where his Neapolitan period ends, at least as far as we know.

---

[22]Letter MF/N5 of March 19, 1938 in Recami (1987).

[23]Probably such "errands" could be run neither the day following this one, a lesson day, nor the day after, as this was a holiday for the anniversary of the foundation of fascism.

[24]Letter MC/N, *loc. cit.*

[25]Letter MF/N6 of March 25, 1938 in Recami (1987).

[26]This and other details we will see later are not taken from Majorana's letters, but from contemporary documents at the Ministry of Internal Affairs, kept in the Central State Archive in Rome (see Recami 1987).

[27]Letter MC/P of March 26, 1938 in Recami (1987).

[28]Testimony reported in Recami (1987).

The Storm

Antonio Carrelli soon got the first signs of Ettore Majorana's disappearance. This is what he wrote to the chancellor of the University of Naples on Wednesday, March 30:

> On Saturday March 26, at 11 in the morning, I received an urgent telegram from my friend and colleague prof. Ettore Majorana, tenured professor of theoretical physics at this university, which goes as follows: "Don't worry. A letter will follow. Majorana". I could not figure out this message, I asked around and understood that in the morning he had not given his lesson. The telegram was from Palermo.

> With the mail delivery of 2 p.m. I received a letter with an earlier date, this time from Naples, where he manifested suicidal intentions. Therefore, I assumed that the following day's urgent telegram from Palermo was meant to reassure me, giving me evidence that nothing had happened. And indeed on Sunday morning I received a priority mail from Palermo where he was saying that the bad thoughts had disappeared and that he would soon be back.

> But sadly the following Monday he did not show up at the Institute.[29]

Worried by this absence, on Monday, March 28, Carrelli called Enrico Fermi in Rome—Amaldi was there[30]—, and read Majorana's letter to him[31]:

> Dear Carrelli,

> I have come to an inevitable decision. You cannot trace a single hint of selfishness in it, but I do understand the problems that my sudden disappearance may cause to you and the students. I beg your pardon for this too, but most of all because I have disappointed you, your true friendship, and the fondness you have shown me these months. Please, remember me to all those I have met and appreciated at your Institute, especially to Sciuti, all of whom I will cherish in my heart at least until 11 tonight, and possibly even after that.

> E. Majorana

Fermi contacted the Majoranas at once. As Ettore's sister, Maria, recalls[32]:

> All at once a telephone call arrived, I cannot remember if it was Fermi or someone else, who asked if Ettore was in Rome with us. "No, he's in Naples... Why?..."

News of his disappearance spread quickly among his friends, too; Gastone Piqué tells us that[33]:

> Enrico Volterra called me up and told me: "Do you know that Majorana has disappeared?" [...] So, I remember I telephoned Amaldi and his brother Luciano, who both confirmed.

---

[29]See document D/ME6 reported in Recami (1987).

[30]See Amaldi's interview in S. Ponz de Leon, *Speciale News: Majorana*, television programme broadcast in 1987 by the Italian TV channel *Canale Cinque*.

[31]Letter MC/N, *loc. cit.*

[32]Interview with Maria Majorana in Ponz de Leon, television programme, *loc. cit.*

[33]Interview with Gastone Piqué in Ponz de Leon, television programme, *loc. cit.*

According to Maria Majorana's information, his brothers Luciano and Salvatore[34] went to Naples the following day, Tuesday March 29, and immediately got in touch with the director Carrelli, and together they went to the Hotel Bologna where Ettore was staying. Once in the room, they immediately found a letter "to my family", which left no doubts:

> I have one desire: do not wear black. If you feel the need to follow the practice, you may carry some sign of mourning, but for no more than three days. After that remember me in your hearts, if you can, and forgive me.[35]

Shocked by what they were reading, and even more puzzled by what Carrelli had told them about the letters and telegrams, Salvatore and Luciano began personally to carry out an exhaustive search which would last for months, in both Naples and Palermo. In the meanwhile, Carrelli announced the disappearance to the Police Commissioner in Naples, and not later than Thursday, March 31, the latter informed the Chief Constable (Senator Arturo Bocchini) in Rome. He immediately alerted all the police headquarters in Italy with instructions to open a missing persons inquiry. But unfortunately, these searches gave no results. Not even Mussolini's personal interest, pushed for by Ettore's mother the following July (Recami 1987), and supported by H.E. Enrico Fermi's intercession, would bear any fruit.

The storm that hit the Majoranas, however, did not pass quickly, as it was fed by hopeful statements (not always reported in official papers) that were gathered, directly or indirectly, by the family itself. And these statements have continued right up until today (for a detailed analysis see Part IV).

First of all, the "tickets mystery" at the *Tirrenia* Shipping Company, which carried passengers between Naples and Palermo. At first it seemed that there was no trace of the ticket receipts for the Naples-Palermo crossing (and return) that Majorana was assumed to have made. However, *Tirrenia* then told the Majoranas that they had found both tickets, and Salvatore Majorana gave this information to Chief Constable Bocchini on April 18, and it was filed away appropriately. Unfortunately, we cannot tell whether *Tirrenia*'s alleged discovery—which was never actually exhibited anyway—happened before or after the statement by Vittorio Strazzeri, mentioned earlier.

There were some alleged sightings of Ettore in Naples, and the family itself, actively carrying out enquiries, gave these varying degrees of credit. Hence, the Prior of the *Chiesa del Gesù Nuovo*, when shown a picture of Majorana, thought he recognised an agitated young man who had came to him at the end of March or the

---

[34]Strangely, in all the documents we have in our possession, and recorded in Recami (1987), only one brother is mentioned. In one of these documents he is identified as Salvatore, to whom Ettore sent his last letter from Naples on March 19. On the other hand, in Amaldi's biography, though still speaking of only one brother, this happened to be Luciano (who was in close contact with Amaldi in the 1960s when they archived Majorana's scientific manuscripts at *Domus Galilaeana* in Pisa). However, Recami's reconstruction in which both brothers came to Naples sounds perfectly plausible.

[35]Letter MF/N6, *loc. cit.*

beginning of April. He had asked to "try the religious way of life" but, as soon as he realised how difficult it would be to remain there for any long period, left without further explanation. And on April 12 another young man was received at the *Convento di S. Pasquale* in Portici, right outside Naples, asking to be admitted to the holy order of the convent, and he, too, was believed to be Majorana; but once again, the young man went away, after receiving a predictable refusal. Finally, a nurse who used to know Ettore claimed to have recognised him at the beginning of April while walking in the centre of Naples.

Clearly, the raging storm of uncontrolled news and sightings from so many different sources was now at its peak.

Part III
A Legacy from the Grand Inquisitor

Chapter 6
The Mystery of the Missing Papers

Naples, "a city where a splendid school of mathematics once flourished, but whose only link with physics was its regret for the disappearance of Ettore Majorana...".[1]

The Director's Office

It is Saturday May 13, 1950. Edoardo Amaldi arrives in Naples on the 1:20 pm train from Rome, with his wife Ginestra, to give the lecture "Accelerating machines for elementary particles". Reaching the Institute of Physics, he engages in a conversation with the director Carrelli who has kindly invited him some time previously. In the director's office there is one of his assistants, Lina Rescigno, a senator's daughter, who is asked to leave the room: clearly the two colleagues want to be left alone. However, even though she leaves the room, the prying assistant is still able to overhear the topic of discussion on the other side of the door: *Ettore Majorana*. Twelve years have gone by, and the storm caused by his disappearance seems to have eased off, but interest in the theoretical physicist, a friend of Amaldi's and a young colleague of Carrelli's, is still alive. Naturally, the interest of the two experimental physicists is strictly scientific.

Among the many that happened before and afterwards, this incident would not be worthy of attention had it not been remembered, quite deliberately, a few years ago by one of professor Majorana's students. And the story, inevitably narrated with some inaccuracies given the long time elapsed,[2] was probably told to Gilda Senatore[3] by another of Carrelli's assistants, Elio Tartaglione. The episode of the meeting between Carrelli and Amaldi does not actually seem to have been about

---

[1] E.R. Caianiello, quoted in (Marinaro and Scarpetta 1996).
[2] It was already quoted by B. Preziosi some time ago: see (Preziosi 1998).
[3] Most of the information about G. Senatore, referred to in this section, was collected during Esposito's interview, mentioned above.

© Springer International Publishing AG 2017
S. Esposito, *Ettore Majorana*, Springer Biographies,
DOI 10.1007/978-3-319-54319-2_6

Majorana's fate, but rather about the fate of his lecture notes on theoretical physics. A distinctively scientific interest, as we were saying. Perhaps.

To understand the whole question, we need to go back to March 25, 1938, when, as already described, Majorana went unexpectedly to the institute. On that day, as a matter of fact, no course lesson had been scheduled, but some students were meeting to go through the latest topics. Majorana must have known about these meetings, if on that March morning he came on purpose to meet a student of his.

At about 11 am, in that long dark corridor on the ground floor of the institute in *Via Tari*, a voice was heard:[4] "Miss Senatore...". The student, alone in a small room where she was writing, came out and found professor Majorana preparing to hand her a closed folder. "Here you are, take these papers, these notes... We'll talk about that later". The professor then walked quickly away, leaving his young student puzzled, but before going out he turned to her and said once again: "We'll talk about that later...". This was the last time Majorana crossed the threshold of the Institute of Experimental Physics.

Though rather surprised, Senatore gave little thought to this incident, because the following day, Saturday, a course lesson was scheduled, and she was sure that they would talk about it then. But the next day there was no lesson, and neither was there on the following days, and when it became clear that the course was not to be resumed, the student kept the folder with scrupulous care, waiting for the professor's next move: "We'll talk about that later...", he had said.

Its content was already familiar to the students of the theoretical physics course; it was the lecture notes prepared by Majorana himself to help his unfortunate students. One of his last thoughts before taking leave was therefore for his students. It says much about the physicist's sincere interest in teaching. The professor might go, but his lessons *had to* stay. And that is the way it went, according to Ettore's plan, who may have tried to foresee the later moves of the minor roles in these events.

In his "farewell" letter to Carrelli, written few hours after giving Miss Senatore the notes folder, Majorana remembered more or less everybody he had met at the institute, but there was no particular mention for that female student. On the other hand, he made a particular reference to another of his students: "Please, remember me to all those I have met and appreciated at your Institute, especially to Sciuti, all of whom I will cherish in my heart...".[5] Sciuti was the only man among Ettore's students, and the only one who occasionally asked for explanations. Sciuti may have talked about his "passion" for Physics with Majorana, and so he remembered him "especially" in his letter to Carrelli. But for the girl he had given his lesson notes to, there was no particular mention. Was this purely fortuitous?

De facto, when the police began to investigate his disappearance, the young Sciuti was questioned without, of course, getting useful results.

---

[4]What follows was remembered by G. Senatore in her intervention at the Department of Physical Sciences of the University of Naples "Federico II" in March 1998.

[5]Letter MC/N, *loc. cit.*

Moreover, the interest in Majorana's lecture notes arose very quickly, if it is true, as claimed by another female student, Nada Minghetti (Russo 1997), that the director Carrelli asked the students about them just a few days after Ettore's disappearance. Ettore may have hinted to his friend Antonio that he was writing up some notes for the students, and the ever alert Carrelli may well have been struck by the possibility that they contain some useful clues for the investigation. An interest, this time, that was not purely scientific.

The Sidekick, the Assistant, and the Co-ed

Meanwhile, the co-ed Senatore[6] kept the notes safely at home in Cava de' Tirreni, at least until the end of 1938. As a matter of fact, after the suspension of the theoretical physics course, she had started to cut down on her visits to the institute, due to health issues; she only resumed a regular attendance after the summer break. But before the break, in parallel with Majorana's course, Senatore had started the preparation for her dissertation and professor Carrelli had assigned her to his first assistant Maione.

Alfredo Maione[7] was 35 at the time (he was born in October 1902 in Sant'Anastasia, a small village near Naples) and had been contributing to the activities of the Institute of Physics in Naples for several years. He was in fact one of the first to graduate with Carrelli, as soon as the latter had returned to Naples to direct the Institute of Experimental Physics. He had graduated in physics during the academic year 1933–34, but already before that he had been appointed temporary assistant from January 1 to June 30, 1933. He had subsequently been made the director's assistant, and after a certain time, in the academic year 1937–38, he became Carrelli's *aiuto* (sidekick), a role he fulfilled until October 15, 1940. For his graduation dissertation he had dealt with magnetic phenomena in certain substances, one of the classic topics studied at the institute before Carrelli's arrival. But when Carrelli became director, Maione's research was also directed towards molecular spectroscopy, although little was published (and, most likely, very little developed either).

This is why Senatore's dissertation concerned spectroscopic research (fluorescence in quinine salts), and she was helped initially by Maione. But, just before Majorana's arrival in Naples, on December 29, 1937, Maione was given a teaching position in physics at the Medical School for the academic year 1937–38 and the following three years. The director Carrelli decided to assign the young student to another of his assistants, Francesco Cennamo, who would help her with the thesis.

---

[6]The following specific information, regarding Senatore and Cennamo, comes directly from Senatore herself, as quoted in the interview mentioned above.

[7]The information here is taken mainly from Maione's personal file at the Archive of the University of Naples and in the yearbooks of the university. Likewise, later, for Cennamo.

This is indeed what happened, after the summer break of 1938; and, as we will see, this change proved to be crucial as far as we are concerned here.

Francesco Cennamo was born in Naples in July 1910. After a short period (1936–37) in temporary appointment, he soon became Carrelli's assistant in 1937–38. Then in 1939–40, when Maione moved out, he became his *aiuto*. He always worked side by side with Carrelli, both in their scientific activities —mainly Raman spectroscopy of liquids—and in their teaching. After the war, he left the Institute of Experimental Physics for the Medical School, after a short period in the Faculty of Veterinary Medicine and the Aviation Academy, where he taught experimental physics in 1959–60. During these years, he published university handbooks and scientific papers with Carrelli, but his research activity came to an end when he was granted tenure.

When he was entrusted with Senatore, just as she was about to graduate, "Ciccio" Cennamo (as he was nicknamed by his friends) was 28 and so just four years older than his student. Handsome in the Rudolph Valentino style, he looked a sort of Don Juan to the girls, who were not indifferent to his charm. And so it was that several of the co-eds fell for the handsome Francesco, including Nella Altieri, whose wealthy family from Salerno invited him over more than once. But this romance was not to be.

Ciccio actually fell in love with the beautiful Gilda Senatore, and after a year helping her out in her studies, over the period September–October 1939, he proposed to Gilda in front of her family. In October 1941, after two years of engagement, they finally got married.

In the meantime, on December 1, 1939, Senatore graduated in physics (just one year later than the other four "innocent souls") and, on the same day, as she remembers, Carrelli hired her as his assistant. Oddly enough, in contrast to Majorana's other students, there is no official record of this appointment in the university archives, nor in the university annual report from that period. However, in the archives, there is a statement from Carrelli in response to a request for a certificate by Senatore, dated September 2, 1944; and it does certify that the young physicist was appointed assistant at the Institute of Physics from December 1, 1939 to November 30, 1940. If that is indeed true, it was probably in an unofficial way and upon the request of her maternal uncle Giovanni Pisapia, an acquaintance of Carrelli's. And anyway, such an appointment was surely also suspended on her fiancé's request, for he would not have approved of the beautiful Gilda working at the institute. Moreover, she did not resume her job because, in 1941, soon after their marriage and when she was expecting their first child, Francesco Cennamo was recalled for duty and had to leave for the war in Albania.

Unfortunately, these personal events are not unrelated to the fate of the folder that Majorana had left on the day of his disappearance.

After Senatore had established a close relationship with Cennamo, perhaps toward the end of 1938 or at the beginning of 1939, the student let the assistant see Majorana's notes, and asked his advice on what to do with them. In fact, he held onto them for some time. But later, without informing Senatore, Cennamo took them to Carrelli, perhaps to discuss some particular topic. However, when he asked to have the folder back, Carrelli refused, for he had in fact been made the consignee for all of Majorana's belongings in Naples.

After the war, once the tumultuous events which had driven Majorana's notes out of mind had finally died down, Mrs. Cennamo began to think of them again, and now and then would ask her husband if he remembered where the precious folder was stored. Francesco Cennamo was fully aware that his wife was jealous of the intimate relationship she had established with her professor, so to avoid her disappointment in finding out what had really happened to the notes, he cooked up a perfectly plausible story. While the Americans were advancing towards Naples, the Senatores' home in Cava de' Tirreni had been hit by a howitzer, probably fired by the Italian Navy, and on that occasion a wooden trunk containing books and other things had been destroyed. Majorana's papers might well have been in that trunk, whence they would have been destroyed during the war.

This is what Senatore believed for forty years, but her husband Cennamo did not forget the episode. Some years after his death (in December 1985), another of Carrelli's former assistants went to see Cennamo's widow: he had been explicitly asked by Francesco Cennamo himself while still living to tell her the truth about the notes, as described above.

War and other tragedies may conceal so many different events for a certain time, but the story we have just told is only one example from the Majorana file. And now other actors are knocking at the door.

The Folder Recovered

What actually happened to the folder containing Majorana's papers? As we have seen, at about the beginning of 1939, it was in Carrelli's hands. However, it is unlikely that he found in it anything relevant for the purposes of the inquiry, if he, a diligent man, did not inform either the police or the Majorana family. As a matter of fact, at the time of the disappearance, his contacts with the family were mediated by Fermi, a close friend of his. But now Fermi was no longer in Italy, as he had left for the USA at the end of 1938, after accepting the Nobel Prize in Stockholm. Actually, only Amaldi had stayed in Rome to continue his research. And, in the second half of the 1960s, Amaldi himself, together with Ettore's brother Luciano, would deliver what must have been in the folder previously given to Senatore to the Domus Galileana in Pisa. In retrospect, all the clues seem to indicate that it was Carrelli himself who handed Majorana's papers over to Amaldi. But when?

One mystery leads to another, and the fact that Miss Rescigno was *required* to leave the director's room on May 13, 1950, so that Amaldi could talk to Carrelli about Majorana does seem a bit suspicious. Might Amaldi have been in Naples with the express purpose of picking up the folder? Somebody believed this. But it is not so.

Carrelli had regularly attended the Institute of Physics in Rome since the 1920s, and must certainly have met Amaldi many times before 1950, not to mention the fact that Amaldi had played an important role in bringing Italian physics back into the limelight after the war, while Carrelli had taken on several responsibilities in this regard, to which we shall return later. Therefore there was never really any need for Amaldi to come to Naples to retrieve his friend's papers.

The reason why Amaldi came to the institute in Naples is, on the contrary, much less mysterious, and in fact of a purely scientific nature; we learn about it in the letters between the two physicists.[8] On January 11, 1950 Carrelli wrote to Amaldi:

> Dearest Amaldi,
>
> I am writing to you on behalf of Cennamo who is carrying out some research in the field of X-ray diffraction in liquids. We have got your results, but it is not possible to have the description of Debye's photographic method, because the key paper is in a publication we do not have. Can you help us to fill this gap? [...] I apologise for bothering you. Please pay my respects to your wife (we have done what was possible for her cousin... although it may not be much...) and best regards to you.

The tone of this letter, as in many others, is clearly affable and suggests a long-term friendship between the two physicists. Amaldi's answer was not long in coming. Indeed, two days later he writes:

> Dear Carrelli,
>
> I am sending you the extract which includes a picture of the device. Unfortunately, there are no plans, but if the picture is inadequate I can give Cennamo any explanation he needs.

Further explanations were in fact needed, and they absolutely required Amaldi to be present in person. Carrelli's official invitation is dated April 18, 1950:

> Dear Amaldi,
>
> I ask you once again to give a lecture here on a topic of your choice. We would like you to come to Naples in the first half of May. Let us know the exact dates and the topic as soon as you can, so that we can welcome you in the best possible way.

And the answer, the following May 8:

> Dear Carrelli,
>
> all being well, I will be in Naples on Saturday. I am leaving on the 1.20 pm train.

The rest has already been told at the beginning of this chapter. Amaldi's visit, though not important for the story of Majorana's folder, was not at all vain for the researcher Cennamo, who would later publish six scientific papers on X-ray diffraction by liquids.

However, we learn that Amaldi knew about Majorana's lecture notes and also that he did not have them, from an enigmatic letter written by Gilberto Bernardini to Amaldi on December 2, 1964:[9]

---

[8]The letters we are referring to are kept at the Amaldi Archive in the Department of Physics at the Sapienza University of Rome.

[9]This letter can be found in (Preziosi 1998).

Dear Edoardo,

here are the pages of Majorana's lectures. I finally found again while clearing out in Geneva. These are just some of them. I reckon Giovannino Gentile has the others, or else they were never written; in fact, I have a blurred recollection from long ago that makes me think that Giovannino gave me them to help me understand Dirac [in fact, Dirac's book]. I have read them again and of course, even now, it is not difficult to spot the echo of his intellect.

This was the letter accompanying the manuscripts of Majorana's lectures, which would later be taken to Pisa by Amaldi. It is enigmatic because of its explicit statements: Bernardini was looking after Majorana's papers, which Gentile had given to him. So two new players enter the scene.

At first glance, this seems to make good sense. Giovannino Gentile was one of Ettore's closest friends, as we have seen, and they shared a passion for theoretical physics, so he was a fair candidate to inherit the theoretical physics notes from the lectures Majorana gave in Naples. Even the fact that Bernardini eventually inherited the notes does not look mysterious: Gilberto and Giovannino had been close friends since university times, when they both attended the Scuola Normale in Pisa (they both graduated in 1927). The theoretical physicist Gentile had lent his notes to the experimental physicist Bernardini to help him get a deeper understanding of quantum physics, as it was presented in the famous book by P.A.M. Dirac (1930).

So everything may be clear, except for the fact that Giovannino Gentile died from blood poisoning (an incurable disease in a time without penicillin) on March 30, 1942, just four years after the disappearance of his friend Ettore. Considering that Carrelli received the folder only at the beginning of 1939, and that afterwards Italy was to go to war, distracting attention from scientific problems, it is easy to see that the span of time for the notes to pass from Carrelli to Gentile is extraordinarily short, particularly if we suppose there was a further intermediate step. Actually, although the two physicists knew each other—Carrelli was on the committee with Persico and Fermi to evaluate the requests for the *libera docenza* (lectureship) in theoretical physics, which Gentile obtained on October 21, 1931—we have no clues that their acquaintance went beyond an ordinary scientific working relationship (which, incidentally, involved quite different areas). It is thus quite unlikely that Carrelli would unthinkingly hand over Majorana's notes directly to Gentile.[10]

But the possible explanations are not exhausted yet.

"Please remember me [...] especially to Sciuti" Ettore wrote to his director just before disappearing. These words might be the origin of another little story, also quite credible, that circulated for a while in the academic environment, and was supported at the beginning by a certain number of witnesses. When presenting the printed version of Majorana's lectures in 1987, the editor wrote: "We want to

---

[10]Besides there is no trace of a possible visit to Naples by Gentile (who was teaching in Milan at the time, as mentioned above) when his friend Ettore disappeared. In any case, such a visit, which should logically have taken place right after the disappearance, would have been totally useless if the idea was to collect the notes, had it not happened at least a year after his disappearance, when Carrelli became the custodian of Majorana's folder.

confirm here that these lectures were given to Bernardini by Sebastiano Sciuti, a student from Majorana's course". And, about ten years later, in a slightly modified version: "Some days before the 25th, Sciuti had received some typewritten pages (concerning the first lectures) by Majorana, and about twenty years later, he gave them to Gilberto Bernardini". This little story has kept going for several years, naïvely sidestepping many "useless" passages to solve the above-mentioned problems.

Actually, after some ups and downs, Sciuti finally managed to fulfil his wish of contributing to "cutting edge" physics with his own research at the Physics Institute in Rome. After graduating with honours in Naples in December 1938, he was soon taken on by Carrelli as his *assistente incaricato* (assistant) for experimental physics from March 29, 1939 to the following August 31. The day after this date he was called for military service, and served as a lieutenant in the Reserve Corps of Engineers at the *Istituto Militare Superiore delle Trasmissioni* (Higher Military Institute of Communications) in Rome until November 30, 1944. Once the war was over, after another short period working at the Naples Institute, from December 1, 1944 until October 31, 1945, he eventually became assistant in Rome in February 1947, and continued his academic career there until his retirement in 1987.

So, as we were saying, the alternative version of what happened to Majorana's papers rested on quite solid grounds. But it is nevertheless totally unfounded: there is no trace of any typewritten document containing the notes of Majorana's lectures, and Sciuti himself finally admitted that he had never owned any papers by Majorana. This little story actually came out *before* the statements by Senatore and her entourage became known, as reported above. Still, it is strange that it circulated for so long.

But this is not the end of the surprises.

The Missing Papers

During the inquiries into the missing Ettore Majorana, which actively involved the director of the Institute of Physics as well, no one ever gave much importance to the possible statements from the few students of the theoretical physics course; they were not expected to know of any interesting and useful details for the purposes of the investigation. With the exception of Sebastiano Sciuti, whom Majorana deliberately referred to in his letter to Carrelli, and who was actually questioned by the police. But he *really* had no information to offer, in contrast to Gilda Senatore, as we have seen. The other physics students, like Sciuti, would make no major contribution to the inquiries.

But what about the "mathematical" guests in Majorana's course? They might not have had any direct contact with the professor, but Ettore knew they were attending his course, and it is strange that after 65 years one of them would supply other pieces for the complex puzzle we are trying to put together.

In September 2004, four years had gone by since Eugenio Moreno's death, when one of his children, Cesare, arrived on the scene to recount what his family had treasured as their father's memories. Like Senatore, he had proudly boasted that he had followed the great missing physicist's course on theoretical physics. Like Senatore, he had jealously held onto a bequest of his: a faithful copy of all the lecture notes prepared by Majorana. At first glance, this may not look like a big achievement to anyone but Eugenio, because we already have Majorana's *master copies*, even though through a somewhat tortuous path. What other information might come out of the *Moreno papers*, apart perhaps from things of scientific interest? But no careful inquiry should ever set aside any element without due consideration, so even this document needs to be examined.

Moreno's copy, as a matter of fact, also records the notes of six lessons which were not in Majorana's original collection (Preziosi 1998). Senatore's or Bernardini's words come to mind here: "they are just a part of it". And a significant part was missing: the master copies record ten lessons, as well as the notes for the introductory lecture (which are not in the Moreno papers).

Tracking the folder Majorana left, whose contents were certainly of great interest to those who managed to get their hands on it, is a much more delicate matter than described so far.

Firstly, if we compare what is in the Moreno papers and in Majorana's master copy, we notice that the pages of the missing lectures are not related to the first lessons, nor to the last, that is, those that Majorana gave just before the course was suspended. Specifically, lecture no. 7 and those from nos. 10 to 14 are missing. The last four deal with a whole subject area, namely Einstein's theory of relativity (Esposito 2006b). Given the position and the topics treated, there can be no doubt that the loss of Majorana's papers is not accidental.

So *who* removed those papers and *why*?

Bernardini's letter to Amaldi, already quoted, seems to offer a simple explanation:

> I reckon Giovannino Gentile has the others, or else they were never written; in fact, I have a blurred recollection from long ago that makes me think that Giovannino gave me them to help me understand Dirac [in fact, Dirac's book].

So here we have a simple explanation that accounts for Gentile keeping the missing papers, but it is not particularly credible. In fact, in the master copy that Gentile is supposed to have given Bernardini, Majorana just barely introduced quantum mechanics, and did not fully develop the subject because of the untimely suspension of the course, so the notes would have been of little help in understanding Dirac's advanced book (Dirac 1930).

Setting aside the theoretical physicist Gentile, there are only a few suspects left who could really have removed Majorana's papers, given the complex itinerary followed by the folder. The experimental physicist Amaldi, who took Majorana's papers to Pisa, had no reason to keep the missing notes and, as a matter of fact, in his scientific and institutional work there is no trace of anything that could be attributed to them. The same is even more true for Cennamo and Senatore (the latter went on to a career outside physics) who also had Majorana's folder in their hands when they were younger.

At least for the time being, the mystery remains.

However, the careful detective should not neglect the role played here by Eugenio Moreno. Although it would be completely unreasonable to think he had any direct involvement, the precise details of how he came to copy down the notes (and above all, when he did it) may help to solve the puzzle.

It should be pointed out at once that the living witnesses of Majorana's course, basically Sciuti and Senatore, both agree on the presence of don Savino Coronato, but have no recollection of Moreno's participation in the lessons, contrary to Moreno's statements to his children and colleagues. If this is true, the notes must have been copied much later, even though this would seriously complicate the situation. For one thing, it would mean that several others must have been involved, such as Caccioppoli. However, this can be ruled out. The Moreno papers are rich in dates and precious historical information, duly verified, and they were in fact the source used here to reconstruct Majorana's stay in Naples. They clearly show that the person who wrote the papers actually attended the lectures, and that the copy may well have been made before Majorana's disappearance. On the other hand, it will still be useful to examine Moreno's affairs in some detail, so let us do that now.

The Adventures of Eugenio M.

Eugenio Moreno[11] was born in Naples on February 16, 1910, the son of Cesare and Sofia Trotta, in their home in Corso S. Giovanni a Teduccio. Until his *Liceo Scientifico* diploma (A levels), which he obtained in 1929, the same year as Majorana was graduating, the young Eugenio was home-schooled and never frequented any public places because his parents were afraid he might catch some kind of infection and die, a not unreasonable concern in Naples in those days. As a result, he grew up to be very shy, and this temperament, which sometimes made him almost invisible to others, remained with him throughout his life, making him similar in this respect to Majorana.

So the first public institution in which he studied was the University of Naples, where he enrolled for the degree course in mathematics at the end of 1929. Through to the very first months of 1932 he studied with regularity, although rather slowly, due perhaps to his personality, but on March 10 he had to attend a training course for reserve officer cadets. Clearly, as he was behind with his studies, he could not ask for a further deferment of his military service, as had been the case for the previous two years, and would be for the following three. He went back to his

---

[11]The information quoted here is partly taken from the interview of Cesare Moreno by Antonino Drago and the writer on September 16, 2004, and partly from Eugenio Moreno's personal file in the archives at the University of Naples.

mathematical studies in 1933, and at this time he made friends with Federico Cafiero and Carlo Ciliberto, two other mathematics students, younger than Eugenio, who would remain friends long after the war. After his first period in the army, he continued with his studies until 1936, when on November 20 he was admitted to the Academy for Reserve Officer Cadets of the 9th Artillery Regiment in Potenza. He was appointed officer candidate on June 10, 1937, and assigned to the 4th Artillery Regiment, where he became Reserve Second Lieutenant the following October 5.

Discharged on December 31, he went back to university again, probably attending Majorana's course on theoretical physics which began a few days after his return. However, this break did not last long, and from October 11 to 25, 1939, he was called back on duty for "instruction" at the headquarters for resting troops of the 10th Artillery Regiment. The preparation of his dissertation was further postponed when, on May 23, 1940, Moreno was again called back on duty (in the Naples military district) and finally graduated in mathematics on the following June 17. After another leave on August 7, Eugenio barely had time to say goodbye to his relatives and friends, and his young fiancé Concetta Lombardi (born in 1924), when he was called up to serve in a war zone, on February 7, 1941.

He distinguished himself in action and after some months was promoted lieutenant, during the worrying wait to see whether he would be called up for duty on the front line. The terrible news came soon afterwards, on January 31, 1942, when he was moved to the 51st Artillery Regiment of the "Siena" Infantry Division, deployed in Bari. Meanwhile, like many other unfortunate fellows on the front, he decided to marry his fiancé "by proxy", and the "wedding at-a-distance" was officially registered on January 10, 1942. Exactly two months later, he set sail for the island of Crete with his regiment to fight on the Greek front, and he arrived at Agios Nikolaos eleven days afterwards. The following April 7 he reached his final destination at Military Base no. 24.

After more than eight months at war, he finally got a special 30-day leave for Christmas 1942, and came back to Italy, flying to Lecce on December 23. He spent a few quiet days far from the front and, once his leave was over, had to return to his military base. However, this time the Neapolitan soldier was luckier, and his return to the battle field was delayed. He left on January 31, 1943, for Greece, arrived on February 12 in Athens, which was not a war zone, and stayed there for about a month, only returning to his regiment on the following March 16.

After a few more months of war, just after the armistice between Italy and the Anglo-American Allied Forces on September 8, 1943, Moreno was captured by the Germans in Crete on September 20 and interned in a concentration camp for the military, where he remained until May 8, 1945. Released by the Germans, Eugenio was then held by the Allied Forces until the following October 13, when he was finally able to return to Italy, and billeted in Rome. Here, after a couple of months' leave, he was put on indefinite leave from the Rome military zone and the terrible war episode of his life could finally be considered to have come to an end. In the meantime, he had been twice decorated with the military cross (*Croci al Merito di Guerra*) in 1950. His health was deeply affected, but little by little Eugenio recovered from this dreadful experience, first by reuniting with his wife Concetta, whom he had married years before by proxy. After some months they had their first child, Cesare Giusto (after Saint Justus, patron saint of Trieste, where the baby had been conceived). He would be followed over the years by other four children: Vincenzo, Angelo, Sofia, and Anna.

Having returned to civilian life in Naples, Moreno and his family had to struggle in a context of economic restrictions and poverty, a common experience among those who had survived the war. He went back to the newly-founded Institute of Mathematics at the university,[12] where he found his old friends again. But things were not quite as he had expected. Cafiero, Ciliberto, and Moreno should all have had to join the army after graduation, perhaps spurred on by Cafiero himself, but the worst fate had been Moreno's. Indeed, when he got back, he found out that Cafiero (and then also Ciliberto) had become an assistant for principles of mathematics, while he had not even had the opportunity to start an academic career as he had been busy fighting the war. But this was not the moment to get into a disagreement with his friends, who helped him as best they could to fit in well in the university. Moreover, Caccioppoli, besides being his former professor, had also become his confidant, and they shared a passion for music, especially piano.

Right after the war, the Institute of Mathematics was in a disastrous situation, and the possibility of employing young volunteers was surely viewed positively. During and after the Allied occupation, in 1944, the premises of the Institute were occupied by the US Military Police and for a long time all lessons were held in the classrooms of the Institute of Experimental Physics, which remained available. Then, after the departure of the Allied troops, the situation got even worse: the classrooms and the offices of the various Cabinets of the Institute of Mathematics were half-destroyed and the library books were piled up on the ground in a large room which did not even have a proper floor. Thanks to the active collaboration of don Savino Coronato and Federico Cafiero, the "reconstruction" of the institute eventually began to move in the right direction, but more "young blood" was still needed.

On December 18, 1945, the tenured professor of advanced mathematics, Carlo Miranda, a close collaborator and friend of Caccioppoli's, wrote to the chancellor Nebbia to ask for a position as *voluntary* assistant for the war veteran Eugenio Moreno. This was promptly accepted and he began on the following January 1 (renewed then until October 31, 1960). Of course the financial situation did not improve, and Moreno tried to find other solutions. He thus entered the selection to teach mathematics in junior high school and got his teaching qualification in 1948. However, he did not obtain a teaching post. Then Moreno had the idea of supporting Miranda's strong commitment to reconstruct the library of the Institute of Mathematics (which lacked permanent staff at the time), and asked for a position, at

---

[12]The Institute of Mathematics was founded on September 29, 1944, when, during a meeting of the Faculty of Science chaired by Carrelli, they bravely decided to merge the institutes of algebraic analysis, infinitesimal calculus, advanced calculus, analytical geometry, descriptive geometry, advanced geometry, rational mechanics, and mathematical physics together with the mathematical seminar into one Institute of Mathematics.

least as librarian. But, despite pressure from Miranda, who also wrote in his favour to the chancellor on May 4, 1949, and the intervention of the Chief Secretary to the Minister of State Education, the position was never established and Moreno was never employed. Luckily, however, on October 1, 1949, he was finally appointed as a teacher of mathematics at the "Macedonio Melloni" Junior High School in Portici, where he taught until September 30, 1968. So his situation finally improved: he continued to be actively involved in research, and above all in teaching at the Institute of Mathematics. In the following years, he first became special assistant for principles of mathematics at the Faculty of Science on November 1, 1960, and then a full assistant on October 1, 1968. He worked at the institute until November 1, 1975, when he retired.

During his long service at the university, Eugenio Moreno often spoke with his colleagues and friends, and also with his children, about his prestigious experience at Majorana's course on theoretical physics, and of course the related notes. However, no one paid much attention to this, perhaps because they considered it to be of little importance. Anyway, Moreno went on studying those notes and always carefully examined any news about Majorana that cropped up in newspapers. The "treasure" he was guarding, however, would have to wait until four years after his death before finally being recovered with all its precious information.

Chapter 7
Fortunes and Misfortunes of a Famous Director

A Devastating Fire

On 11 April 1952, the police headquarters in Naples sent a letter to the university asking for "the place of birth and full personal details" of "Ettore Majorana son of Fabio, born 5 August 1906", who, "in 1938, was a tenured professor at this university". The police enquiries of fourteen years earlier to elucidate the mysterious disappearance of the physicist from Catania had come to nothing, as we will see in detail later, so the time had come to dismiss the Majorana case. It may of course seem utterly strange that, after so many years and, above all, after the end of the war, the police who had benefited so much from the collaboration of the Majorana family, would spontaneously dig out the case of the missing physicist. However, the explicit request, which shows that they at least knew that the missing person had a tenure at the University of Naples, was made purely to establish Majorana's personal details so that the police could subsequently file the case.

The *Ufficio del Personale* (human resources) at the university was quick to answer the official request and, after a few inquiries, the *negative* answer came on April 16: "The file has been destroyed. There is nothing in the archives here". But there was more: "Professor Ettore Majorana was a tenured professor of theoretical physics, appointed by law according to article 81 of Law 31.8.1933 no. 1592—for his high repute". This last statement, which seems to contradict what had just been said (and it was true that the file had been destroyed), reveals an interesting side to the story: the law mentioned was *not the right one*. The law on higher education referred to (which was actually approved by Royal Decree on 31 August 1933) came *before* the legislative decree of June 20, 1935, which had been used to appoint Majorana "for his high repute". This latter decree was cited explicitly in the

© Springer International Publishing AG 2017
S. Esposito, *Ettore Majorana*, Springer Biographies,
DOI 10.1007/978-3-319-54319-2_7

Minister Bottai's official letter to Majorana and, more importantly, in the Minister's note to the chancellor of the University of Naples on 5 November 1937:[1]

> We inform Your Lordship that, by application of article 8 of the Royal Legislative Decree of June 20, 1935 no. 1071, Professor Ettore Majorana is appointed, from November 16, 1937, to the tenure of theoretical physics at this Faculty of Science, for the high repute of the particular expertise he has achieved in the field of said discipline.
>
> We would therefore ask Your Lordship to give notice of this to the relevant faculty and to Professor Majorana.
>
> signed Giustini

On the one hand, the mistake shows that there really was no official document left in the archives, and on the other it suggests that, at least at the university, some kind of fact-finding survey had been made. This survey showed that Majorana held a theoretical physics tenure and that this position was given "for his high repute"—neither piece of information was in the official document of the Police Headquarters requesting the information.

The official answer, written by the chancellor Ernesto Pontieri to the police headquarters in Naples the following April 29, is also interesting. He basically wrote that they "were unable to supply the information requested [...] as all the documents in the archives had been destroyed", without mentioning what they had ascertained,[2] which was thus considered useless.

The destruction of Majorana's personal file—and other documents in the university archives—was one of the calamities brought upon the University of Naples by the Second World War. It also had some consequences for the affairs we are dealing with here, so this story is worth telling.

The day after the Badoglio Proclamation on 8 September 1943, announcing the armistice between Italy and the Allies, when the German troops were about to leave Naples, a German soldier was killed by an unknown assailant near the university building in *Corso Umberto I*. The German command suspected (for no good reason) that the fatal shot had come from university premises, so they ordered those premises to be set on fire in retaliation. Hence, on September 12, the chancellor's office, the administrative offices, some department secretaries' offices, the classrooms of the Department of Law and Humanities, the Institute of Industrial Chemistry, and even the well-stocked library of the *Società Reale di Scienze, Lettere e Arti* were all burned down. Unfortunately this barbaric deed was not the only one the Germans inflicted on the city of Naples. There was also the fire that

---

[1]The original manuscript of this document is providentially kept in the Central State Archive in Rome, in Ettore Majorana's personal file at the Ministry of Education (*Dir. Gen. Istr. Sup. - Fascicoli Personale Insegnante e Amministrativo, II° Versamento - 2° Serie - B93*).

[2]This information was handwritten on the letter from the *Questura* (police headquarters) and then registered by the *Ufficio del Personale* (human resources). The corresponding document was therefore an internal note of the university, which was shown to the chancellor to help him write his answer; then it was archived in Majorana's personal file, specifically recreated on this occasion. No other document appears in this file.

destroyed the priceless (from a historical and cultural point of view) documents kept in the State Archives, in the heart of the city.

The 1943 fire thus destroyed all the documents relating to Majorana's appointment and almost all the official documents (up to 1943) relating to the other characters involved in our story. However, this has only had a small effect on the task of putting together the facts we have presented here. As a matter of fact, the availability of alternative sources[3] has in part bypassed the obstacle. Furthermore, the academic administration had to reconstruct the careers of university employees up to 12 September 1943, so they themselves proceeded to fill in the gaps. On 8 March 1951, the chancellor issued a decree that set up a committee to find and check information about the careers of university employees, and this information was then written down and "officialised" in the person's official records (*Foglio di Notizie*). So, for instance, Sciuti's record sheet (written on 24 March 1954) was filled in on the basis of a letter from Carrelli to the chancellor (dated October 1944) and a surviving chancellor's decree of March 1939. For the Minghetti file (written on 8 October 1951), a sworn statement by Carrelli in front of a notary was required.

Senatore's file was a different case again. She had requested a certificate in September 1944, that is, before the institution of the above-mentioned board. As no official document was available, the certificate was released by the university's human resources department, based on a simple statement by the director Carrelli. He declared that Senatore had been hired at the Institute of Experimental Physics "as *assistente incaricata* from 1 December 1939 to 30 November 1940", receiving a special mention as assistant to the director, who "with deep regret saw her leaving the institute for family reasons". So this is the only documented information testifying to her appointment (besides Senatore's own statement), but there is no trace of it in the university report for that academic year, while there is for Sciuti (for a shorter period of time), Minghetti, and others.[4]

Rumours About a Powerful Man

The director Carrelli played a role, both indirectly and directly, that was anything but secondary in accurately handing down information for our story, as we have seen in the previous pages. Therefore, we need to pause for a moment to fully understand how he actually fulfilled this role. So here is a quick sketch of Antonio Carrelli.

---

[3]Mainly annual reports of the University of Naples and oral statements by the same people.

[4]Actually, Carrelli's statement for Sciuti is also (partially) incorrect, as it brings forward the actual dates when the assistant was hired by three months, giving 1 January to 1 September 1939, instead of 29 March to 31 August 1939, as confirmed by the chancellor's statement.

Those who take on this task, looking at statements by those who knew him (mainly former students or collaborators) will soon notice, surprisingly enough, that pretty much all the statements stress the *negative* features of Carrelli's personality, with the sole exception of his personal secretary, Mrs. Cutillo, who thought very highly of "her" director. Together with accounts of true events and circumstances, there have often been quite sensible but sometimes odd interpretations which have passed on an image of Carrelli that serves little purpose as far as our story is concerned.

> Carrelli loved dramatic gestures. He would walk through the door of the great hall in *Via Tari*, where the general physics lessons were given to physicists, mathematicians, and engineers alike, like a prima donna […]. He would not sign the students' exam transcripts personally, but left the janitor Mario Esposito to do that for him: Esposito collected all the transcripts and stamped them in purple, after being tipped.[5]

And again:

> During the course, Carrelli slavishly followed his own book (full of errors, both in form and content; rumour had it that he was not the author of the book, but that it was written by one of his assistants) and at the exam he demanded that students use its exact words.

As regards his collaborators, the tone of the comments is the same: "Carrelli treated his assistants badly, often yelling at them; […] everybody was terrified". These statements clash somewhat with Majorana's impression, which we have already hinted at, but they should probably not be taken out of the peculiar context they were formed in and which they refer to. Here we are talking about the social context of the 1950s and 1960s and the ensuing youth protests, on the one hand, and on the other, Carrelli's prominent, sometimes political, roles during those years, not to mention a possible change in Carrelli's personality after the death of his wife Eleonora Laliccia in the 1950s.

This situation is also sadly confirmed by the fact that none of his Neapolitan colleagues remembered his work in the appropriate places after his death on 25 November 1980, although his contributions should in no way be underestimated. The only tribute to the director of the Naples Institute was duly held one year later by the *Accademia Nazionale dei Lincei*, which Carrelli had presided over for many years (Radicati di Brozolo 1981). However, this question is not strictly important for the reconstruction of the Majorana case, so we shall try now to move on, and refer only to the facts that are relevant to the matter at hand.

---

[5] Aldo Covello's statement, given to the writer on 7 July 2005. From now on, only this statement is reported, as an example. In fact, statements by other former students and collaborators express similar sentiments.

Antonio Carrelli

Antonio Carrelli was born in Naples on 1 July 1900, the son of Raffaele and Silvia
Scardaccione. After attending the best high school in Naples, *Liceo Vittorio
Emanuele*, he enrolled at the University of Naples at the age of 17 to study physics.
However, he was soon obliged to interrupt his studies because of the First World
War. He was called up very young and, entering the war after training as a reserve
officer cadet, he left the conflict as a second lieutenant in the Grenadiers. This did
not prevent him from graduating in physics when he was only 21, under the
supervision of the director of the Institute of Physics Michele Cantone. He then
became Cantone's assistant and would succeed him as director after his death. In
1922, he entered the selection for a grant to study abroad from the Faculty of
Sciences at the University of Rome, and he was the runner-up, after Fermi. The
report by the selection committee, made up of Salvatore Pincherle (chairman), Luigi
Berzolari, Michele Cantone, Orso M. Corbino, and Nicola Parravano, clearly
expresses the highest consideration for Carrelli's scientific work.

> He has dealt mainly with optical experiments, and in particular has examined absorption
> and fluorescence phenomena. He studies the absorption of iodine solutions using a pro-
> cedure that can pinpoint the maxima, and accurately determines their dispersion law, by
> verifying the presence of points of inflection in the curves corresponding to the absorbing
> areas. Of particular interest is the study of the fluorescence of various colouring substances,
> where the author shows that in dispersion curves there are anomalies near both absorption
> and emission areas. A theoretical study of the Brucke effect is worth a mention. We also
> find proof of physical and mathematical skills on a range of questions relating to spectral
> analysis and atomic structure.
>
> Considering that this candidate only graduated a little more than a year ago, a period of
> study abroad would certainly be useful to develop his striking aptitude for scientific
> research.[6]

In 1924 he was granted the lectureship (*libera docenza*) in experimental physics,
but never presented any coursework at the University of Naples. That same year,
instead, he was appointed (in Naples) as professor (*professore incaricato*) of theoretical physics and taught topics such as optics and atomic spectroscopy, which
were cutting-edge for the time, well before Fermi became a tenured professor in the
first teaching post specifically created for these subjects in Rome in 1926. These
were also the years of his connections with the Institute of Physics in Rome, which
would last until long after Fermi's departure for the USA. It is curious to note, for
instance, that in 1928–29 the Rome Institute (twice) funded Carrelli's work and
scientific research.[7] In Fermi's own papers, and in those of other foreign authors,

---

[6]See the *Bollettino del Ministero dell'Educazione Nazionale* (Part II: Administrative Acts), year 50
(1923), p. 802.

[7]This information can be obtained from the note for the accounting period 1928–29 at the Institute
of Physics in the University of Rome. These accounts are kept at the Museum of Physics in the
Sapienza University of Rome.

Carrelli's spectroscopy research is cited more than once, attesting to a keen interest in such accurate studies.

Concerning the education of the young physicist, his stays in Berlin were decisive. The German city was one of the most dynamic centres for the study of the rising quantum mechanics. There he was able to attend lectures by Max Planck, Albert Einstein, Max von Laue, and Walter Nernst. As recalled by one of Carrelli's colleagues, not only did this experience orient all his future research

> [...] towards the field of optical spectroscopy – which was to remain for years the centre of his experimental activity – but it also woke in him a strong interest for the theoretical problems of relativity and quantum mechanics, which were in those years the focus of the famous Berlin weekly seminar. Still many years later, in our conversations in Naples, Carrelli would remember almost in awe those lively discussions at the Berlin seminar, attended by the greatest physicists of the day [...] (Radicati di Brozolo 1981).

In Berlin he also began an active collaboration with Peter Pringsheim on several questions of molecular physics, whose results were published in the best scientific journals of the day (and also in the well-known and prestigious *Handbuch der Physik*).

The university tenure came in 1930 after a series of curious "coincidences". In that year there was a selection for a temporary professor of experimental physics at the University of Catania. The committee appointed by the ministry to evaluate the candidates met in Rome on 29 October 1930, and was chaired by Quirino Majorana. The other members were Antonino Lo Surdo, Giuseppe Grossi Cristaldi, Michele La Rosa (from the University of Catania), and Alessandro Amerio. Twelve candidates entered the selection, including, apart from Carrelli, Franco Rasetti and Gleb Wataghin. Here is the committee's appraisal of Carrelli:

> Antonio Carrelli graduated in physics in Naples in 1921 with honours and began as assistant of mineralogy then physics at the Royal University of Naples. He won a development scholarship from the Scientific and Technical Committee; lecturer since 1924; member of the Accademia di Scienze Fisiche e Matematiche of Naples since 1928; he won the "Sella" award in 1927 and was among the three winners of the contest for best paper on quantum theory, organised by the Accademia Pontificia dei Nuovi Lincei. Since 1924 he has taught theoretical physics at the Royal University of Naples.

> He is presenting 45 publications: 29 theoretical and 16 experimental, among which 4 in collaboration.

> Some of the first are reviews of the literature on important modern questions, such as splitting of spectral lines by electric fields, new statistical concepts, and quantum theory. The original notes deal with: the speed of radiant energy in a fluorescent medium and dispersion in the case of broad asymmetric bands; calculation of the diffusion coefficient of electromagnetic radiation; the compound photoelectric effect; everything based on the most recent concepts. He finds a double period in the number of magnetons of the elements between calcium and zinc. Important is the group of papers on wave mechanics, where he often reaches remarkable results, when calculating the number of complexions related to the three different statistics, and when he studies the Compton effect or the theory of induced fluorescence. Some deal with problems of diffusion, the Tyndall effect, and the Raman effect. For latter, Carrelli derives the theory, by demonstrating that the effect can be considered as a diffuse radiation. He also proves that both the magnetic effect of a circular plate

rotating around its axis and crossed by a radial current, and the rotation of the plane of polarisation produced by a rotating dispersive medium are negligible. The experimental part is often linked to some theoretical research, as we can see in his studies of the Tyndall effect and the Raman effect, which confirm his theoretical predictions. Commendable also are the studies on Barlow's wheel, on dispersion and absorption by iodine solutions; he finds that, in fluorescent solutions, the ordinary and rotatory dispersion of light occur at an anomalous rate in absorption, while in emission the anomaly is only in ordinary dispersion. If fluorescence is excited by polarised light, it is partially polarised; the Tyndall effect excited by polarised light gives rise to totally or partially polarised light, depending on the case, and the emission is always of different intensities in the various directions transverse to the exciting ray. Commendable, finally, are his studies on the broadening of spectral lines. The complexity of this scientific activity, most remarkable if one considers that it has been developed in just nine years, clearly demonstrates this candidate's sound theoretical knowledge regarding the most modern questions of physics. As the remarkable experimental investigations are often linked to theoretical developments, and sometimes stem from them, Carrelli stands out as an expert on physics, especially when addressing theoretical questions. Given also his teaching experience, the Committee is unanimous in considering him to be highly suitable for a tenure in experimental physics.[8]

The decision on who should be selected was somewhat difficult:

The Committee, having examined and discussed the candidates' credentials, concludes that Carrelli and Rasetti stand out quite clearly. All the other candidates considered suitable follow at some distance.

And among the final three, Carrelli and Rasetti ended up co-winners. However, a later vote decided that Rasetti should be appointed to the tenure, and thus move to Catania. However, as already mentioned, Corbino did not want such an effective experimental collaborator for Fermi to leave "his" Rome group. So he worked hard, and managed to establish a position in spectroscopy in Rome, which Rasetti accepted. This left the experimental physics teaching post in Catania free, and Carrelli was asked to go there. Once at the Institute of Physics in Catania, he also became its director and totally reorganized it. The following three years' activity is well documented in the report by the evaluation committee for Carrelli's advancement[9] to tenured professor of experimental physics; the board was made up of Q. Majorana, E. Fermi, and Laureto Tieri, and they met on 9 April 1934, in Rome:

Professor Carrelli, chosen for the chair of experimental physics, was appointed temporary professor of experimental physics at the Royal University of Catania on December 1, 1930.

On November 1, 1932, he was called to direct the Institute of Physics at the Royal University of Naples. During his three years of activity, he spent the first two years in Catania and the third in Naples. During his second year, he went to Holland on a grant from

---

[8]See the *Bollettino del Ministero dell'Educazione Nazionale* (Part II: Administrative Acts), year 58 (1931), vol. II, p. 1604.

[9]Winning the selection for a university chair meant being hired as *professore straordinario* (*adjunct* professor or, in other words, *not permanent*), followed by a three-year probationary period. At the end, a special board would assess the adjunct professor's work and, if the response was positive, he would be attributed a permanent tenure (full professor).

the Reale Accademia d'Italia. Professor Carrelli's activity in this special period has been excellent, both in teaching and in the reorganisation of the Institutes of Physics in Catania and Naples, as can be seen from the reports of the Faculties of Sciences of the Royal University of Catania and Naples. [...]

Professor Carrelli's publications in these three years confirm his sound theoretical knowledge and his commendable experimental aptitude, the latter being particularly clear in his papers on the Raman effect.

Hence, the Board unanimously considers professor Carrelli worthy of promotion and suggests that H.E. appoint him to the tenure.[10]

Carrelli's zeal in reorganising the Naples Physics Institute was praiseworthy:

In 1932 Carrelli was asked to succeed his teacher Cantone as director of the Institute of Physics at the University of Naples, and soon after he devoted himself with great enthusiasm to modernising its technical equipment and research directions, determined to make Naples one of the most active centres of Italian physics. This was why he strove for Ettore Majorana to take the position in theoretical physics, established thanks to his direct intervention; though very young, Majorana had already shown his genius and obtained results of the highest importance. Unfortunately, Carrelli's dream of developing, around Majorana, an active school of theoretical physics was brutally interrupted by the sudden and mysterious disappearance of the young scientist. Carrelli spoke about Majorana, even many years later when I arrived in Naples,[11] with emotional admiration and tender regret (Radicati di Brozolo 1981).

Even before his promotion to tenure, Carrelli's national prestige had begun to grow, as reflected in the roles he played and the positions he was appointed to. For example in the 1930s he was appointed to many selection committees, sometimes associated with Fermi. In 1936 he was elected corresponding member (*socio corrispondente*) of the prestigious *Accademia Nazionale dei Lincei*, and became effective member (*socio nazionale*) in 1947, in good company with the most distinguished men of learning at the time. Two years later he became chairman of the 4th science section (physics) in the Physical and Mathematical Sciences Group of the Italian Society for the Advancement of Sciences (*Società Italiana per il Progresso delle Scienze*). He was also a member of many other cultural institutions, and was eventually made a *Cavaliere* of the Crown of Italy.

During the Second World War Carrelli, Dean of the Faculty of Sciences since 1940, distinguished himself, together with Caccioppoli and others, in the democratic revival of the University of Naples. A particularly significant episode was the following.[12] During the German occupation of Naples, the well-known chancellor of the University of Naples, Adolfo Omodeo, was stuck in Salerno, and could not fulfil his role, being too far from his place of work. So Carrelli, who was aware of

---

[10]See the *Bollettino del Ministero dell'Educazione Nazionale* (Part II: Administrative Acts), year 61 (1934), vol. II, p. 2378.

[11]Radicati di Brozolo was called by Carrelli to Naples for the tenure of theoretical physics in 1953.

[12]This episode is reported by Carlo Miranda in the meeting of the Faculty of Sciences at the University of Naples on 15 and 22 July 1976, as can be read in the minutes.

the importance of Omodeo's presence in Naples to get the university started again, sailed off to Salerno in his boat to bring him back to Naples.

At the end of the war, the Neapolitan physicist

[...] did his best, not only to renovate what had been destroyed, but also, and I am quoting a famous colleague's words, "to eliminate as far as possible the incrustations of a centuries-old arteriosclerosis". A very open-minded spirit when it came to scientific problems, Carrelli did his best to bring the liveliest minds to Naples from all over Italy. [...]

Meanwhile, Carrelli was busy organizing the recovery of the scientific research carried out in his institute. Due to lack of funds and the decreasing interest in optical spectroscopy, Carrelli decided to redirect research towards the study of the properties of condensed matter. [...] Two other lines of research were developed by Carrelli and his skilled collaborators during the last years of his scientific activity. The first was the design and development, with the active participation of Professor Porreca, of the first Italian microtron for accelerating electrons up to energies of 2.5 MeV, a machine which was later used by his collaborators to study radiation damage in solids. The knowledge gained by the Neapolitan physicists through the building of this machine proved to be very useful when, later on, they had the idea of using a microtron as an injector for the Accelerator Laboratories at Frascati (Radicati di Brozolo 1981).

In 1945, due perhaps also to his "patriotic" commitment during World War II, Carrelli was appointed vice-president of the RAI (the Italian Television Broadcasting Company), on the suggestion of the Naples National Liberation Committee, and he subsequently became president from 1954 to 1960. During this period, the television company was totally modernised: a first television network was set up, then a second, then a frequency modulation radio network, and the first experiments in colour television transmission got started. Since then, together with his job as teacher and researcher (which he continued until the 1970s, though at a slower pace), Carrelli maintained a high level of institutional activity. He became president of other institutions, such as the Committee for Physics of the C.N.R., the Technical Council of the Post and Telecommunication company, and the Ugo Bordoni Foundation for research in telecommunications, but also of important companies involved in regenerating the country's industry, such as Salmoiraghi, Società Meridionale di Elettricità, and Microlambda. In addition, he was an influential member of the Science Committee of the UNESCO Commission and vice-president of Euratom in Brussels in 1965.

In November 1975 Antonio Carrelli retired from the University of Naples, which later made him professor emeritus, the perfect ending to a brilliant institutional and scientific career.[13]

---

[13]Carrelli's last paper dates back to 1970, but his scientific production continued essentially uninterrupted until then, and even in his last period, he produced many single-author papers.

The Blue Colour of the Heavens

Before getting back to facts strictly related to Majorana (and Carrelli), let us pause for a moment and deal with a topic which may seem rather far from our present purpose: why, during the day, does the sky look blue, while it goes a reddish hue during sunrise and sunset?

It is common experience that a beam of light travelling through a uniform medium, such as a short tract of water or air, will propagate undisturbed through it and in a straight line. Things change if, on the contrary, there are some spatial or temporal variations in the properties of the medium (and in particular, the electromagnetic properties), as for example when sunlight travels through vast tracts of the Earth's atmosphere. In this case, there is a phenomenon known as light scattering: part of the light energy is deflected from the initial direction of propagation, changing certain characteristics of the scattered light, such as its colour, as compared with the unscattered incident light. The essential features of this phenomenon were already understood around 1500 by Leonardo da Vinci, though only on a qualitative level, in his experiments on the scattering of sunlight passing through wood smoke when observed against a dark background. However, more quantitative observations were only carried out some 350 years later, in 1870, with the work of J. Tyndall. The theoretical framework was then worked out some time later by Lord Rayleigh in his famous book on blue sky and sunset, using J.C. Maxwell's electromagnetic theory of light. He showed that white light (which is well known to be a superposition of all the colours of the rainbow, from red to violet) coming from the sun is scattered by the atoms of the atmosphere in such a way that the least scattered component is red and the most scattered is violet. As a consequence, if we look directly at the sun we perceive it as white, and if we look in a different direction we notice a higher percentage of blue-violet components than in the incident light. Essentially, therefore, the blue colour of the sky, like the red colour of the sun at sunset, is a consequence of the phenomenon of Rayleigh scattering of sunlight in the atmosphere.

Beyond the interest raised by Lord Rayleigh's explanation of the above-mentioned phenomena as an application of Maxwell's theory, physicists continued to research this matter even in the following century, owing to a peculiar feature. As a matter of fact, in his private letters to Rayleigh, Maxwell had already observed that the attenuation of light in the atmosphere would not have been possible if the medium it is made of (air) had been continuous, that is, if it had not had a granular structure. Actually, the problem is closely related to the atomic structure of matter, a widely debated topic in the scientific community between the end of the nineteenth and the beginning of the twentieth century. In particular, the above-mentioned phenomena, and more precisely, the measurements of light attenuation in the atmosphere, may be used (and indeed were used) to estimate how many atoms there are in a given amount of matter under certain conditions, leading to an important *non-chemical* determination of the so-called Avogadro number. No wonder then that some physicists in the first half of the twentieth century engaged in a deeper study of the effects already observed by Tyndall some time previously, both on the experimental side and as regards the detailed theoretical interpretation.

The Papers that Were Never Found

In 1924–25, the young Carrelli carried out some research on the Tyndall phenomenon in the Physical Institute of the Royal University of Naples, under the supervision of his teacher Michele Cantone, who later presented the results of this research at the *Accademia dei Lincei*, to which he belonged (and the results were also published in German). Carrelli sought to understand certain peculiar, almost negligible, effects which had already attracted the attention of Tyndall and Lord Rayleigh, but whose interpretation was still somewhat unclear.[14] In his study, Carrelli analysed some peculiar characteristics of the phenomenon and achieved some interesting results,[15] although they were not of the utmost importance.

What is important to us here, however, is something else. Almost twenty years later, in 1946, Carrelli went back to that same phenomenon (in the meantime he had been dealing with completely different problems) and within the space of a few months (from May to November) published no fewer than four notes (Carrelli 1946) in the *Rendiconti dell'Accademia dei Lincei*. In Carrelli's own words "the purpose of this work is to develop the theory and bring to light many facts not yet clearly elucidated". Indeed, in these four notes, the author reported some very detailed studies in which he personally carried out experimental investigations, while also developing the theory required to interpret his own observations and those already known.

In the light of what already said, it should come as no surprise that Carrelli reached such interesting results, even if on a topic which was no longer in the spotlight of modern physics, being rather a phenomenon of "classical" physics. In any case, we must take into account the fact that the means available at the Institute of Physics in Naples were rather limited, and experimental work, even in advanced fields (the studies carried out in Naples focused mainly on atomic and molecular spectroscopy), could only be performed if it was easy to implement. The research carried out by Carrelli on Tyndall's phenomenon falls precisely into this category.

But it is perhaps strange that the director went back to an old topic of his own accord, so many years after first investigating it, especially since it was no longer an important field either in the restricted circles of research in Naples, or, more

---

[14]Under some circumstances, if we illuminate a turbid medium with natural light, and observe the light scattered in a direction perpendicular to the incident light, we notice that this is richer in blue colours than expected according to Rayleigh's original theory. This effect, called "residual blue" by Tyndall himself, is explained by supposing that light is scattered *twice*, thus increasing the blue component of the scattered beam.

[15]The final paper is in (Carrelli 1925).

significantly, in the broader circles of international research. This circumstance thus hints at some contingent event which must have reawakened Carrelli's interest.

It is now well known that six lessons are missing from the master copies of Majorana's lecture notes on theoretical physics in Naples, but we know what they were about from the Moreno papers. Expert attention has focused mainly on four of these six lessons, which develop Einstein's theory of relativity. As already mentioned, the course on theoretical physics at the University of Naples had been given by Carrelli since 1924, except for the years when he was in Catania and the short period when they were delivered by Majorana. During the long span of time when he gave this course, Carrelli never dealt with the theory of relativity until the year following Majorana's disappearance. That year, as a matter of fact, Carrelli also published his lectures on this topic for his students (Carrelli 1940). The coincidence between these events is certainly surprising, especially if one considers, as pointed out by Senatore and others, that by that time Carrelli was (or had been) in possession of Majorana's original notes. However, from this "coincidence" we cannot reasonably say that Majorana's missing notes were taken by Carrelli. In fact, a comparison between Carrelli's and Majorana's lectures on the theory of relativity seems to exclude any clear dependence of one author on the other.[16] The "coincidence" might even be explained if we admit that Carrelli might have been attracted to the topic independently of the autographed notes, for instance, by direct discussions with Majorana himself when he gave his lessons. In any case, Carrelli was well aware that Majorana had given some lectures on the theory of relativity, because the students had complained to the director about their much too challenging level:

> Majorana explains the theory of relativity in four lessons, and his pupils run to Professor Carrelli and complain because Majorana's course "is too difficult". In the ensuing conversation, Carrelli and Majorana get to know each other better and become friends: "Take it easy, Ettore", Carrelli recommends, "your pupils are not following you" (Zullino 1964).[17]

It is instead more surprising that, in the missing lecture just before the beginning of the presentation of the theory of relativity, Majorana dealt with some consequences of Maxwell's electromagnetic theory, and in particular, Rayleigh's theory of the blue sky. What Majorana presented to his students had nothing to do with the particular effects mentioned above, which were studied by Carrelli. Majorana discussed the classical theory as developed by Rayleigh, although as usual he dealt with it in his own way. However, this further "coincidence" does not seem to be totally unrelated to Carrelli's revived interest in the Tyndall effect if, as he states in

---

[16]As was his habit, Carrelli did not deal with the theory of relativity at an advanced level, while Majorana did. And this even though Carrelli himself had already carried out some theoretical research on particular relativistic topics. See, for instance, the paper in (Carrelli 1928).

[17]It is worth noticing that the testimony reported by the journalist Zullino, which clearly comes from Carrelli himself, is the only one referring to Majorana's lectures on the theory of relativity (and also the one with the correct number of lessons, only four), and it comes about forty years prior to the discovery of the Moreno papers.

his paper, the research (even the theoretical research) on this particular phenomenon had started well before 1946, and in any case in the heat of World War II.

The other missing lecture, the first of the group of six, is unrelated to either Rayleigh's theory or Einstein's relativity, but deals with a particular aspect of spectroscopy, the main field of investigation at the Naples Institute of Physics.

Carrelli's scientific attitude, always of the utmost rigour and intellectual honesty, as attested throughout his many years of research activity, excludes any possibility of a misappropriation of Majorana's papers. But the three "coincidences" described here seem to point to a temporary loan—for study and research purposes—to the person who was in any case the official custodian of Majorana's belongings in Naples. It does look as though the lecture notes and other possible papers were not given immediately to the Majorana family along with Ettore's other personal effects, but then those manuscripts were given to Carrelli long after Ettore's brothers arrived in Naples following their relative's disappearance; and also after Fermi's departure for the States, at the end of 1938, bearing in mind that it was through his friend Fermi in Rome that Carrelli communicated with the Majoranas. It is then plausible that the subsequent loss of the six lectures, clearly kept apart from the others, can be put down to the idea that they were being used for study purposes, and indeed Carrelli almost suggests this himself. Indeed, in the first of his four notes "on the polarisation of the light from the sky", he explicitly states that:

> This work which is now coming to light was mostly carried out in the year 1943–44 when, because of the war, the Institute of Physics at the Royal University of Naples was actually closed and the writer was in Meta di Sorrento.

Although the hypothesis does not stand on firm foundations, it would not be strange or difficult to accept that the fate of Majorana's notes could be closely linked to the tragic events of war, recalled several times here, and to Carrelli's escape to Meta di Sorrento, where he carried out his research on the Tyndall effect.

The dissemination of news relating to Majorana has proved once again to follow tortuous paths. And beware! The "coincidences" are not over by any means.

Part IV
Investigation of a Disappearance

Chapter 8
Before March 26

"Have you seen them?—Our colleagues Majorana E. and Segrè have disappeared from the corridors of the *Scuola di Applicazione*. A weird chap has been noticed around Termini station. Rumour has it that it was Majorana E. Anyway, we are anxiously waiting to hear from the two of them".[1]

A Surprising Coincidence

The reconstruction of the investigations carried out into Majorana's disappearance from Naples in 1938 should necessarily start with primary documents and the related witness statements. This has already been done by others, and accurately enough. But then it would not be very useful to keep thinking in the same way as the investigators of the time (the police, relatives, and friends), since that brought no definite results regarding Ettore Majorana's disappearance. Of course, if we want to report the facts and circumstances related to that mysterious disappearance with any accuracy, we cannot disregard those documents and the statements collected by the investigators at the time. However, in our analysis we shall proceed *independently* from those investigators, presenting some facts, and some reasonable hypotheses derived from them, that *did not* come out in the 1938 investigations, but only recently.

The starting point is an event which apparently has nothing to do with Majorana's disappearance. As already reported, on Tuesday, January 18, Ettore met Giuseppe Occhialini who, on his way back from Brazil, had decided to pay a short visit to the director of the Institute of Physics in Naples. Occhialini's dramatic

---

[1] This humorous note was posted in the corridors of the *Scuola di Applicazione per gli Ingegneri* in Rome by a close friend of Majorana's, Gastone Piqué (and subsequently kept by him; see F. and D. Dubini, *La scomparsa di Ettore Majorana*, cit.), after Majorana and Segrè dropped their engineering studies, while Piqué and other friends stayed on.

© Springer International Publishing AG 2017
S. Esposito, *Ettore Majorana*, Springer Biographies,
DOI 10.1007/978-3-319-54319-2_8

statement came out only in 1990 and it is generally interpreted as an early indication of Ettore's alleged suicide. Occhialini puts it like this:

> "It's strange", I told him, "that having wanted to meet you for so long when living in Florence, this meeting has only been possible by going to Sao Paulo and back!"

> Answer: "You have made it just in time, because if you had come a few weeks later, you wouldn't have met me".

> I sensed, I understood at once, from the tone of his voice, from his gaze, what he was talking about. He was talking about something I knew very well. So I said: "This… this answer of yours makes you even more interesting to my eyes, because I too, since I was 18, have been repeating words like these to myself … I have been walking the same path…"

> And he: "Dear Occhialini, there are people who talk about it, and people who do it. That's why I repeat that if you had arrived a few weeks later, you wouldn't have seen me" (Russo 1997).

But let us linger on another aspect of this, equally intriguing.

The vessel Occhialini disembarked from in Naples, which had sailed from South America and would continue to Trieste, was the ship *Oceania* of the *Società di Navigazione Italia* (Italia Line) based in Genoa, the same shipping company that owned the transatlantic *Rex* and the future *Andrea Doria*. It operated (together with its twin *Neptunia*) a regular service between Trieste and Naples in Italy and South America.[2] It seems likely that this *insignificant* detail may have come out in the conversation between Occhialini, Carrelli, and Majorana.

The surprising "coincidence" here is that the same ship *Oceania*, after another voyage to South America (starting on February 4 from Trieste and February 6 from Naples), set out again from Trieste to Buenos Aires, after a stopover in Naples on Saturday March 26, the very day that Majorana disappeared…

We may also note that, among the few other books in Majorana's very limited personal library, there was a copy of the 1937 Nautical Almanac,[3] the last one issued before his disappearance. The almanac mainly covered the Italian and foreign naval fleets in service that year, but a few pages were devoted to Italian passenger ships in civilian service. It is somewhat curious to observe that, in the almanac Majorana procured at the beginning of 1938 (the one relating to the year 1937), the only cruise ship pictured was precisely the *Oceania*…

The mention here of Buenos Aires will surely remind the attentive reader of other testimonies, ones that point to Argentina (Recami 1987), which we shall discuss in due time. For the moment, the clue given indirectly by Occhialini's testimony suggests that we should further investigate this idea, following a lead overlooked by the investigators of the "Majorana case".

---

[2]The information reported from here on can easily be found in the newspapers of the time, mainly *Il Mattino* in Naples and *Il Giornale di Sicilia* in Palermo.

[3]This information was given to the writer by Ettore Majorana jr, the son of our protagonist's brother.

Leaving the Port of Naples

The ferry Majorana probably took on that Friday in March to go to Palermo was the *Città di Palermo* of the Tirrenia Shipping Company. As might be expected of one of the biggest ports in Italy, Naples saw the passage of many other ships, on many different routes. However, some of these were not passenger ships, but only carried cargo, and henceforth we shall only consider those ships also carrying passengers.

On Friday March 25, 1938, the ships leaving the port of Naples were as follows:

- *Castellon* (German), to Catania;
- *Rastrello*, to Torre Annunziata;
- *Casaregis*, to Mogadishu;
- *Città di Tripoli*, to Benghazi;
- *Leonardo da Vinci*, to Port Said and Mogadishu;
- *Città di Palermo*, to Palermo.

Since all the information collected by Ettore's relatives at the time, together with the letter Carrelli received from Palermo, suggest that Majorana caught the postal ferry to Palermo on Friday 25, it may seem pointless to linger on the departures from the port of Naples on the days following that Friday. However, as we were saying, we would like to proceed independently from what has already been done, if only to reconstruct the most general picture possible of the situations Majorana may have encountered during those few days. So let us consider the ships leaving Naples on Saturday, March 26:

- *Oceania*, coming from Trieste and heading to Buenos Aires;
- *Yasukuni Maru* (Japanese), coming from London and Marseille and heading to Port Said and Yokohama;
- *Puccini*, to Livorno;
- *Cilicia*, to Messina and Istanbul;
- *Alcora* (English), to Messina;
- *Santa Marina Salina*, to Messina;
- *Città di Tunisi*, to Palermo.

For the following days, we mention only the following ships leaving on Sunday, March 27:

- *Saturnia*, to North America (stopover in Palermo the following day);
- *Hakozaki Maru* (Japanese), to Marseille, Gibraltar, London, Antwerp, Rotterdam;
- *Exeter* (American), to Boston and New York;

and the following leaving Naples on Monday, March 28:

- *Chisone*, to the North Pacific (stopovers in Livorno, Genoa, Marseille);
- *Egeo*, to Rhodes;
- *Excambion* (American), to Alexandria, Jaffa, Haifa, Beirut.

At the time of Majorana's disappearance there was therefore a great deal of traffic in the port of Naples, including ships bound for several foreign destinations, with an impressive 12 ships in 4 days! This information was regularly reported in the Naples newspaper at the time, so it is quite plausible to think that Ettore, who had been keen on ships and naval affairs from an early age, would have been aware of this traffic.

The next departure from Naples to South America was on April 23, on the ship *Neptunia*.

It is also interesting to note that the Neapolitan agent of the *Società di Navigazione Italia*, owner of the ships *Oceania*, *Saturnia*, and *Chisone*, and the one of the *American Export Line*, owner of the *Exeter* and the *Excambion*, were both based in *Piazza G. Bovio* (the former at no. 22, the latter at no. 14), which is about halfway between the Hotel Bologna (in *Via Depretis*) and the University (in *Corso Umberto I*), the route Majorana would have followed every day.

One last "curious" piece of news comes from Palermo's *Giornale di Sicilia* of March 28, 1938:

> This morning, on the mail ship "Città di Palermo" to Tripoli, Lady Chamberlain and her daughter have come to our city and are heading to Libya for a short break. The honoured guests [...] after a short visit to the city, went back on board just before the departure. H.E. Marshal Balbo's wife was also on board.

The mail ship mentioned here is the one Majorana was supposed to have caught on the evening of March 25, though counting on the following trip departing from Naples on Sunday 27. The presence on board of the wives of the English Prime Minister and a fascist party official would certainly have meant much closer attention by the police force, both at the port of Naples and in Palermo; attention not paid, however, to a man who was about to silently disappear. It would be really interesting to know whether Ettore was also aware of this circumstance.

A Matter of Money

Let us now return to January 18, the day of the meeting between Majorana and Occhialini, to investigate other things that may have happened after that meeting.

For this purpose, the first document we have is a letter Ettore wrote to his mother on Saturday, January 22:

> I have just finished the fifth lesson. [...] During next week I will be short of money; so could you ask Luciano to withdraw my share of the bank account and perhaps send me all of it, first taking into account the previous withdrawals and returning the 1000 liras you recently gave me. [...] I think I will come in a few days, but only for few hours, because I have to get a book at Treves and others from home.[4]

---

[4]Letter MF/N2, *loc. cit.*

The two interesting facts mentioned in the letter concern the lessons given at the university and the money matter. Let us pause for a moment on the latter.

Majorana had only recently started his job at the Institute of Physics in Naples (recall that the employee register was filled in on January 20, only two days before the letter quoted here), and so he would not yet have received his salary (in fact his mother had had to lend him 1000 liras). It was therefore no surprise that he would be "short of money", and hence the obvious request for his brother Luciano to withdraw his share of the bank account in Rome. We do not know how much money was involved, but we do have some information about university salaries.

In the Minister Bottai's appointment note, registered at the Court of Auditors on December 4, 1937, we can read:

> With effect from November 16, 1937, prof. Ettore Majorana is appointed, for the high repute of particular expertise he has gained in the field of the studies of theoretical physics, Full Professor of Theoretical Physics at the Faculty of Mathematical, Physical and Natural Sciences of the Royal University of Naples.

> Starting from the same date Professor Majorana is placed at level VI, group A, with a salary of liras 22,000 plus liras 7000 e.f.a. [except following adjustment], respectively reduced to liras 19,871.72 and liras 6322.82 (except following adjustment).[5]

(The quoted figures must of course be read as annual salary, paid on a monthly basis.)

Majorana's appointment for exceptional merits had also brought him some financial advantages; indeed, the "ordinary" winners of the tenure (Wick, Racah, and Gentile) were in the "VII level, group A, with a salary of *liras* 19,000+ *liras* 5200 e.f.a. respectively reduced to *liras* 17,551.99+ *liras* 4801.10 e.f.a.", starting from December 1, 1937.[6] Ettore probably withdrew his first salary (corresponding to two and a half months from November 16 to January) at the end of January from the Naples branch of the Bank of Italy. Anyway, we can be sure of the total amount of money Majorana withdrew before disappearing. Indeed, on August 16, 1938, the chancellor Giunio Salvi of the University of Naples responded as follows to a specific request from the Ministry of Education on July 30 of the same year:

> I can assure you that the last salary Professor Majorana withdrew is the one for last February. From March 27 on, no other remuneration has been paid to the above-mentioned professor. Salaries for March and April have regularly been transferred back to the Treasury.

> Since May 1, the note naming Professor Majorana has no longer been issued.[7]

---

[5]Almost completely reported in Recami (1987); the full text can be read in the Decrees of the Court of Auditors of December 1937, kept in the State Central Archives in Rome.

[6]Decrees of the Court of Auditors of December 1937, *loc. cit.*

[7]See the documents in Ettore Majorana's personal file at the Ministry of Education, *Dir. Gen. Istr. Sup.—Fascicoli Personale Insegnante e Amministrativo, II° Versamento—2° Serie—B93*, kept in the State Central Archives in Rome.

So, taking into account the money his brother Luciano withdrew from the bank,[8] and his expenses for the stay in Naples, it seems fair to say that, at the moment of his disappearance,[9] Majorana would have had about 10,000 liras with him.

Clues from the Neapolitan Lectures

The second interesting fact, in the letter to his family of January 22, is the reference to the lessons Majorana was giving at the Institute of Physics in Naples. The specific and detailed mention of the "fifth" lesson, which he had just given that Saturday, although we only find it in Ettore's letters (and specifically the letters addressed to his family[10]), may be accidental and hence of little importance. The question of the books to be collected in Rome, however, is quite relevant, if we relate it to the fact that, after this fifth lesson and after collecting those books, Majorana started to write the lecture notes for his students. The easiest and most reasonable explanation is that the professor realized that his students were encountering some difficulties in following his advanced lessons of theoretical physics (and his course would undoubtedly have been the most advanced among those followed by the Neapolitan students of the time), so he decided to help them by providing them with some written assistance. One of the books he wanted to get in Rome would have been the Fermi's *Introduzione alla Fisica Atomica*, which we have already mentioned, and which Ettore himself had used as a student ten years before at the university in Rome. Anyway, the fact that he thought about these books, and the notes, during the weekend just after his meeting with Occhialini,[11] may not be accidental. We do not want to put forward here an interpretation that is not based on solid grounds, but it may be useful to bear in mind such an alternative.

In any case, this observation suggests that we should examine the lecture notes written by Majorana, looking for possible clues regarding his disappearance. Furthermore, this is the last written text he left us, and one of the last things he did before disappearing was to hand it over to one of his students. So what information might we extract from the notes of his course on theoretical physics that could be useful to our investigation? We have noted the following things.

In designing any academic course made up of several lessons, the teacher who wants his/her students to grasp the theoretical conceptual unity underlying the given

---

[8]It is curious to note that Ettore's request to his mother—to withdraw all his money from his bank account—would certainly have aroused suspicion had it been made at any time *after* withdrawing his first salary; but this was not the case for the request of January 22.

[9]None of Ettore's relatives testify to any amount of money being left by Ettore in his hotel room or anywhere else.

[10]See the letters reported in Recami (1987).

[11]As far as we are concerned, Ettore would visit his family in Rome only at the weekends, though not regularly, when he had most of his free time, from Saturday afternoon and including the whole of the following Monday.

discipline will naturally make cross-references to lessons before or after the one he/she is giving ("as we have already seen", "as we will notice later on", etc.). One often sees this in textbooks and handbooks. It is perhaps slightly more unusual to find such references explicitly reported in informal notes, written by a professor to help his/her students understand a series of lectures. On the other hand, there is no reason why this should not be the case, so there is no need to be particularly surprised by such a thing.

In Majorana's course notes we do not find frequent cross-references between the different lessons. But the distribution of the cross-references to later lessons that were *never* given is *very peculiar*:

- 2 references in lecture no. 7 of January 27,
- 2 references in lecture no. 8 of January 29,
- 1 reference in lecture no. 10 of February 5,
- 2 references in lecture no. 20 of March 22,
- 2 references in lecture no. 21 of March 24.

Majorana felt the need, intentionally or otherwise, to reassure his audience in writing of the natural development of his course, but referring to lessons which were never given. This happened only on three occasions: the first, right after Occhialini's visit, and the last, on the days before his disappearance. In any case, it is strange that the three occasions correlate, although perhaps *accidentally*, to the passage of the ship *Oceania* in the port of Naples, onward bound for South America (January 18, February 6, March 26).

French Papers

According to the recollections of his family, Ettore "did not like to frequent women":[12] clearly, such a complex personality would not tend to consider or be attracted by such matters. As already noted, in his "farewell" letter to Carrelli he specified: "Please remember me [...] especially to Sciuti".[13] It was reasonable, then, to expect Majorana to give the folder with his notes to his only male student, that is, Sciuti, who was also the only one actively participating in the lessons. Instead, it was given to a woman, and according to what Senatore herself has told us, this was not at all accidental. We do not know the reason for such a choice. The only difference we might note is that, among all of Majorana's students, Senatore, although Italian, was born in Sao Paulo; something Majorana is likely to have become aware of through his loquacious friend Carrelli, who used to go to the theatre with the girl's uncle.[14]

---

[12]See, for instance, the interview with Maria Majorana reported in (Ferrieri and Magnano 1972).
[13]Letter MC/N, *loc. cit.*
[14]Esposito's interview with G. Senatore, *loc. cit.*

So what was in the folder handed to Senatore on that Friday, March 25?

Of course, all the lecture notes on theoretical physics written by Majorana (corresponding to 16 lessons, from no. 6 on), and also a manuscript, not meant for the students but for his personal use, probably referring to the text of a lecture or a general seminar Ettore may have given at the Institute of Physics in Naples, on topics of interest for the director and his collaborators.[15] But Senatore also recalls that "we are still missing a few papers, original handwritten copies, very neatly set down like the others, but which were not part of the lessons already given",[16] papers which unfortunately we know nothing about, either their content or their fate. However, we cannot ignore the fact that, in the folder *now* kept in Pisa and containing both the lecture notes and the manuscript relating to the seminar in Naples, there are three further papers (Esposito 2007) whose content was certainly not intended either for the students or for the physics researchers in Naples. Instead, these contain personal research notes,[17] and in particular, an outline of a theory of quantum electrodynamics, using the mathematical formalism of quantum field theory.

However, what interests us here is something quite different: this document is the only known scientific text written by Majorana in *French*. Though it is not surprising that Ettore knew this language quite well (and German, too), at least as far as technical matters were concerned, it is nevertheless rather intriguing and one would like to know his reasons for writing this. It is reasonable to think that it was not written during or even close to his stay in Naples, but rather around 1933–1934 (Esposito 2007). Speaking of which, it is natural to associate this text with the only other document we know of written by Majorana in French. It is the draft of one of his letters responding to an invitation to a conference:

Dear Sir,

I thank you very much for your invitation to participate in the next congress that will take place in Leningrad. I am glad to accept and to have the opportunity to visit, at the same time, your big and beautiful country. I have also discussed your invitation with Mr. Fermi and Mr. Rossi. Fermi is busy with a series of conferences in America and will not be able to attend. Rossi, on the contrary, has assured me he will gladly accept to go to Russia.[18]

Especially after his stay with Heisenberg in Leipzig, Majorana's scientific work had become known abroad. It is therefore no wonder that he was invited to a conference and, through him, also Fermi and Bruno Rossi. The latter was a young researcher who, though working at the Institute of Physics in Padua at the time, often visited

---

[15]See the introduction in (Majorana 2006) and (Esposito 2006).

[16]See G. Senatore's intervention at the Department of Physical Sciences of the University of Naples "Federico II" in March 1998.

[17]Strictly speaking, we cannot be absolutely sure that the papers we are referring to here were actually in the folder given to Senatore, although there are reasonable indications that this was the case. However, we shall assume this here simply in order to explore the possible consequences for our investigation.

[18]The original in French is reported (letter MX/R1) in Recami (1987).

Fermi's group in Rome; he, too, was appreciated abroad for his accurate experimental work on cosmic ray physics.

Even though the letter carries no date, detailed cross-checks have shown that the Russian conference may have been either the First All-Union Nuclear Conference in 1933 or the International Conference on Theoretical Physics in 1934. The former took place at the Leningrad Physico-Technical Institute from September 24 to 30, 1933, and its main topic was the physics of the atomic nucleus. It brought together many important physicists working in this area, such as I.E. Tamm, V.A. Fock, G. A. Gamov, P.A.M. Dirac, F. Joliot, V.F. Weisskopf, and also a "lad from *Via Panisperna*", Franco Rasetti. It is curious to note that, in his talk, Dirac referred to the topic also dealt with by Majorana in the French text, but from a different point of view. The other Russian conference, though also organised by important Soviet physicists working in Leningrad (for instance A.F. Joffe), was not held in that city, but in the other important centre of Kharkov, in May 1934. This congress, which did not focus solely on nuclear physics, was attended by only a few non-Russian physicists, among whom N. Bohr and L. Rosenfeld. These two knew Majorana well, as did V.F. Weisskopf and R. Peierls, who, the year before the conference, had worked for a long time at the institutes in Kharkov and Leningrad, respectively.

Contrary to what is stated in the letter, Majorana did not attend either of the two conferences, or any other in the Soviet Union, and the same goes for Rossi, but we do not know the reason why.

Anyway, another aspect is interesting for our purposes here, namely, the reason why Fermi could not participate, as he "was busy with a series of conferences in America". As a matter of fact, if the invitation to Majorana concerned the Leningrad conference, at around that time, Fermi was in the United States, where he had been invited to give some lectures at the Ann Arbor summer school, in Michigan, in August 1933, together with Emilio Segrè. However, if we take what Majorana wrote literally, it may be more plausible that the invitation concerned the Kharkov conference. In this case, we know that in the summer of 1934, Fermi went to South America, where he actually held a series of conferences in Buenos Aires and other places.

If this last fact were to be confirmed (like the ones which precede it and are the premises to it), we would once again be forced to point out a curious coincidence between certain facts relating to South America and Majorana's last moves before his disappearance.

Chapter 9
In Search of a Missing Professor

Getting the Search Started

Little or nothing of what has been discussed in the previous chapter came to light in the enquiries carried out by the police and Majorana's relatives just after his disappearance. Though these never brought tangible results, they nevertheless provide, directly or indirectly, very interesting information. We shall therefore examine them here in great detail. Such information is now mainly found in three files on Majorana, kept in the Central State Archive in Rome[1]; most of it has already been disclosed,[2] and in part used in the previous chapter, but it is now extremely interesting to analyse and interpret it in the light of what has been said above.

The enquiries started on Tuesday, 29 March 1938, thanks to Antonio Carrelli and Ettore's brother (Salvatore). The director of the Naples Institute of Physics informed his chancellor, Giunio Salvi, in the following letter:[3]

---

[1]The files are those of the Home Office, PS Series—1939—A1, Envelope no. 51 and Political Police Series, Envelope no. 758, and also the one from the Ministry of Education, *Direzione Generale Istruzione Superiore, Fascicoli Personale Insegnante e Amministrativo, II° Versamento* —*2° Serie*, Envelope no. 93. For the sake of brevity, we shall refer to them henceforth as PS file, Political Police file, and MPI file, respectively.

[2]We owe the discovery and publication of this material to the patience and care of Professor Recami; see Recami (1987).

[3]Letter in the MPI file; document D/ME6 in Recami (1987).

© Springer International Publishing AG 2017
S. Esposito, *Ettore Majorana*, Springer Biographies,
DOI 10.1007/978-3-319-54319-2_9

Royal University of Naples

Naples, March 30, 1938

Confidential

Chancellor,

It pains me to inform you that

on Saturday March 26 at 11 in the morning I received the following urgent telegram from my colleague and friend Professor Ettore Majorana, tenure of Theoretical Physics at this University: "Don't be alarmed. A letter will follow. Majorana." I considered this message totally unintelligible; I asked around at once and learned that he had not given his lesson that morning. The telegram came from Palermo.

With the mail delivery of 2 p.m. I received a letter with a previous date, and from Naples, in which he expressed suicidal thoughts. I then understood that the following day's telegram from Palermo was meant to reassure me, giving me evidence that nothing had happened. And indeed on Sunday morning a priority mail from Palermo arrived, in which it was stated that the bad thoughts had disappeared and he would soon be back.

But unfortunately the following day, Monday, he did not show up, either at the Institute or at the hotel where he was staying. Alarmed by his absence, I informed his family in Rome about the turn of events. Yesterday morning his brother arrived here, and we went together to the Police Commissioner of the city of Naples, asking him to find out from the Police Commissioner of the city of Palermo whether Professor Majorana was still there in some hotel of that city. As this morning I have not yet received any further information, I am informing Y.M. about what is happening, hoping that my colleague just wanted to have some time off, after a reaching a state of exhaustion, or a moment of depression, and that he will soon return to us to pursue his active and intelligent contribution.

Respectfully yours

Signed Antonio Carrelli

Disregarding the obvious information in this document, it is interesting to note that Carrelli, before sounding the alarm about Majorana's disappearance, carried out a few personal inquiries, to be sure about what to do (which he then did, impeccably). So, after receiving the telegram on Saturday 26, he asked around and discovered that his colleague had not given his lesson (and only *after that* did he realise that the telegram came from Palermo); in the afternoon, after receiving the letter of the day before, he realised that the telegram gave him "evidence that nothing had happened". On Monday 28, before informing the Majoranas in Rome (through Fermi), he did not just wait for Ettore to come into the institute, but inquired as to whether he had come back to the Hotel Bologna.

Possible speculations are based on the fact that Majorana decided to disappear not at the beginning or in the middle of a working week, but at the end of it. But independently of these interpretations, the fact is that the police search began only four days after the disappearance. We do not know whether that was accidental, or whether Ettore had actually predicted the way his friend Carrelli would act. It is interesting to note, however, that the very first interpretation he gave of Majorana's disappearance was not that it was due to a "moment of depression", something which could occur very suddenly, but rather that it resulted from a "state of exhaustion", which usually requires longer to manifest itself.

Carrelli's letter was a note for the chancellor of the University of Naples, who received it the following day, Thursday 31, and immediately sent it to the Ministry of Education in Rome:[4]

Royal University of Naples

Naples, March 31, 1938

To the Honourable Minister of Education

General Department of Higher Education
Rome

Subject: Professor Ettore Maiorana

Confidential – Urgent

I hasten to send you, by official duty, the copy of a letter which has arrived today from the Director of the Institute of Physics, regarding Professor Ettore Maiorana.

Unfortunately, we have not yet received any news of him.

I have personally got the Police Commissioner involved in a possible enquiry and I reserve the right to inform this Honourable Ministry of any further news.

The Chancellor

G. Salvi

All the authorities had been alerted, so the search could begin.

First Enquiries

On March 31, in the evening, two days after Carrelli and Ettore's brother had alerted the Police Headquarters in Naples, the first account of what had happened arrived at the General Affairs Division of the Home Office in Rome:[5]

---

[4]Letter in the MPI file; document D/ME6 in Recami (1987).
[5]Document in the PS file; reported in Recami (1987).

Professor Ettore Majorana, of the late Fabio and Dorina Corso, born in Catania on August 5, 1906

Tenure of Physics at the Royal University of Naples.

_____

On March 25 (Friday) he wrote a letter from Naples to the director of the Institute of Physics, Professor Carrelli, where he said that he had to take the inevitable decision of leaving his teaching career. He informed that he was leaving Naples by sea. He left the Hotel Bologna where he was staying at about 5 pm.

On March 26 (Saturday) in a letter, from Palermo, he informed Professor Carrelli that he would be back (perhaps with the same letter) in Naples, where he was supposed to arrive Sunday 27 or Monday 28.

On the same day (Saturday 26), he cabled the Hotel Bologna from Palermo to keep his room, where he had left clothes and papers.

March 31 (Thursday). Until this morning, we have no news of him.

Professor Carrelli saw best to declare his disappearance to the Naples Police Commissioner.

_____

His family have carried out searches with some apprehension: it transpires that no one under this name took the Naples-Palermo or Palermo-Naples steamer.

The young man, acutely misanthropic and not in good health, may have retired to some destination in Palermo, or to some private clinic. He may have left for Tunis. It is unlikely that he has gone anywhere he may be recognized (for example, to Catania).

He has a passport valid in Europe, renewed last July.

Height 1.68 metres – Oval face – Large and lively eyes – Dark hair, dark complexion – Steel -grey overcoat – Dark brown hat.

This communication was probably issued by the Naples Police Headquarters by telephone or "by telegram" (as noted on it), and, *oddly*, written on paper carrying the letterhead of an (unknown[6]) member of the Council of State, and it was registered on this same paper the following day, 1 April 1938.

It is surprising that all the information in this document was quite clearly collected by the missing person's relatives, and not by the police who, in the two days that had gone by, *would not appear* to have carried out their own enquiries. Moreover, the inaccuracies in the report are quite striking. As a matter of fact, in neither of the two letters of March 25 (to Carrelli and to his family[7]) did Majorana state that he would leave Naples "by sea"; this can only be read, of course, in the letter to Carrelli from Palermo on the following day,[8] which also contained the other information reported in the document, except for the allusion, also mistaken, to his return to Naples on

_____

[6]Very likely, the state councillor was Ettore's uncle Oliviero Savini Nicci (husband of Ettore's father's sister Elvira), who intervened personally in the case of his missing nephew (Roncoroni 2011).

[7]Letters MC/N and MF/N6 of 25 March 1938 in Recami (1987).

[8]Letter MC/P of 26 March 1938 in Recami (1987).

"Monday 28". It could well be that the police were not explicitly shown Ettore's letters, but only informed of their content by Carrelli and Salvatore Majorana. Finally, the last part of the report (with the exception of the information about the passport and the physical description) looks like a rather free interpretation that would be unusual for the police of the Kingdom of Italy. Alternatively, it could indicate a certain superficiality on the part of the investigators, with the facile allusion to a "private clinic", because Ettore was "not in good health", or even the unfounded claim that he was en route for Tunis, which could easily have been proved wrong just by checking the departures from the port of Palermo (or Naples).

Anyway, after immediately informing the Home Office in Rome (on the same day, March 31), the Chief Constable sent the Coding Office a telegraphic message for all the police officers of the kingdom—telegram no. 10639:[9]

442 Stop

Please look for the sole purpose of finding him and without letting him know professor of physics of the Royal University of Naples Ettore Majorana of the late Fabio - of Dorina Corso born in Catania August 5, 1906 left Naples without informing family Stop Professor Majorana holds European passport renewed last June or July Stop Physical description Stop Height 1.68 metres – Oval face – Large and lively eyes – Dark hair, dark complexion – Steel-grey overcoat – Dark brown hat Stop In case of finding urgently cable ministry indicating his possible movements Stop

Chief Constable.

Note here that the Chief Constable's main concern was not letting the missing person know of the search they were carrying out, but merely indicating "his possible movements".

The Home Office officials were quick to pick up the information, unfortunately negative, that came from the police headquarters, duly noting on the same despatch "Expired" or "No response" (dated April 3 and 7, May 20 and 22), before filing it for good.

The Political Police Enters the Scene

The enquiries into Majorana's disappearance were not bringing anything relevant for his family, who were obviously very worried. All Ettore's friends and colleagues felt the same, including Giovannino Gentile who, overcoming his humble disposition (Gentile 1942), asked his father, an important figure in fascist Italy, to intervene in person.

Hence, on Saturday April 16, the day before Easter in 1938, Senator Giovanni Gentile wrote an "urgent" letter to "H.E. Senator Arturo Bocchini", chief of the fascist police service:[10]

---

[9]Document in the PS file, reproduced in Recami (1987).

[10]Letter in the Political Police File; document D/PP in Recami (1987).

Senate of the Kingdom

Rome, April 16, 1938

Dear Excellency,

I would beg you to receive Dr Salvatore Majorana, who needs to discuss the case of his unfortunate brother, the missing professor.

From a recent clue it would seem that a new enquiry is necessary, in the convents of Naples and its surroundings, and maybe throughout the whole of central and southern Italy. I would strongly urge you to do this. Over the past few years, professor Majorana has been a major contributor to Italian science. And if we are still in time to save him and bring him back to life, and to science, as we all hope, we must not neglect any lead.

Cordial greetings and best wishes for Easter.

Yours

Giovanni Gentile

This letter was attached to the request for a meeting presented by Ettore's brother on the day after Easter:[11]

HOME OFFICE

Request for meeting

Rome,

April 4, 1938

Mr

Salvatore Majorana

Resident in

Rome

Profession:

Law graduate

ccupation:
Requests to speak to:

H.E. Senator Bocchini

Subject of visit

(specify)

To report on important clues relating to the disappearance of professor Ettore Majorana - Letter from Senator Giovanni Gentile

The request was immediately accepted, and Salvatore Majorana was received at once by Senator Bocchini. We know about the content of that meeting from a note

---

[11]Request in the Political Police File.

written by a ministry official during or, more likely, soon after the meeting with Ettore's brother:[12]

Rome, April 18, 1938

Subject: Professor <u>Ettore Majorana's</u> disappearance (and possible suicide)

Mr. <u>Salvatore Maiorana</u>, brother of Professor Ettore Maiorana, now missing since March 26 of this year, refers to other details ascertained by the relatives themselves.

After carrying out searches in collaboration with the Naples police, in both Naples and Palermo, nothing has been established. Professor Maiorana went from Naples to Palermo with thoughts of suicide (according to letters he left), and therefore was thought to have remained in Palermo. But this theory must now be dismissed, because the return ticket has been found at the offices of "Tirrenia", and also because he was seen at five o'clock, asleep in the steamship cabin, during the return voyage. Then at the beginning of April, he was seen – and recognized – in Naples, between *Palazzo Reale* [Royal Palace] and the *Galleria* [City Gallery], walking up from Santa Lucia [street], by a nurse who knew him and who also saw and guessed the colour of his suit.

Given this information, and since his relatives are now convinced that Professor Maiorana has come back to Naples, they ask for all hotel check-ins in Naples and the surrounding area to be re-examined (Maiorana is spelt with a "j", <u>Majorana</u>, which might have been the reason why the name was not detected in the first round of searches), and they ask the Naples police – who already have the photograph – to intensify the search. Some enquiries could be made to see whether he bought any weapons in Naples from March 27 on.

Once again we cannot help but note that, at least up to this point, the searches were carried out by the missing person's relatives "in collaboration with the Naples police". Anyway, since the last official report, some (supposedly) "new" details have emerged from the enquiries.

First of all, the about-face by Tirrenia, who seemed, strangely enough, to find Majorana's Palermo-Naples ticket, not in the days right after the journey, but several days later. We have already mentioned how this puzzled his friends and family. Note that there are signs of this puzzlement even in the present document: the discovery of the ticket is *also* confirmed because Ettore "was seen at five o'clock in the steamship cabin". We do not know how this information was obtained, but it is a fact that, one way or another, Salvatore Majorana got in touch with Professor Vittorio Strazzeri, a well-known mathematician from the University of Palermo (see next chapter), who claimed to have travelled in the same cabin as Ettore (and a third passenger) on their way back from Palermo to Naples. After their encounter, on May 31, Strazzeri felt the need to write a letter to Salvatore Majorana:[13]

---

[12]Request in the Political Police File; see also Recami (1987). In the top left corner of the document, we read: "Letter of recommendation from Senator Gentile attached".

[13]Statement T/6 in Recami (1987).

Dearest Mr Majorana,

it is my absolute belief that, if the person who travelled with me is your brother, he has not committed suicide, at least not before arriving in Naples.

In fact, when I got up, we were entering the port of Naples, and many passengers were on the deck of the steamship, as it was already broad daylight.

I repeat, in the cabin I did not see any luggage, but what struck me was that maybe a vest or a jacket (or some sort of clothing) had been put on the net hanging over each bed; that struck me because, whenever I am travelling, my main preoccupation is to watch over my wallet and passport.

I do not doubt that the third passenger was called Carlo Price, but I can assure you that he spoke Italian just like us southerners, and what's more, he looked like a shopkeeper or something like that, a person lacking that unconscious refinement of manners deriving from education.

I repeat once more that, if the young man who travelled with me was your brother (I am saying "young" because he was not bald and because I got that impression), he certainly did not commit suicide before the arrival of the steamship in Naples.

Please, kiss your mother's hand on my behalf and pay my respects to your family.

If you receive any news, do keep me informed. I believe that – if they are as good as I hope and expect – they will give me much joy.

Your devoted servant,

                                                                    V. Strazzeri

Palermo, May 31, 1938

P.S.: Forgive me if I dare suggest that you look for your brother in some convent, as happened in the past with not very devout people, I think, in Monte Cassino.

His brother Luciano was already convinced of the improbability of Ettore's suicide, in particular, by jumping from the Palermo to Naples ferry: "a battalion of veterans from Africa was travelling on the steamship; the deck was crowded. They could not have helped noticing a man falling into the sea".[14]

Apparently, then, this letter does not contain anything new or useful which would justify its writing. Here Strazzeri actually just confirms what he had already said to face to Salvatore Majorana, perhaps giving a few more details (and adding a "suggestion" in the end). However, what is striking in the general tone of the letter is that he twice uses a phrase like "if the person who travelled with me is your brother", perhaps indicating an unconscious change of mind by the witness, who may no longer have been quite so sure about his identification.[15]

---

[14]See the interview with Luciano Majorana in S. Nicolosi, "Lo scienziato Majorana non fu rapito" in the Italian magazine *Visto*, February 1959.

[15]For the record, we would also like to point out that, according to the recollections of Ettore's relatives, other statements were made by people who had seen Majorana on board the ferry,

The news in the note of April 18, which certainly would be more valuable as it comes from someone who knew the missing person, is that Ettore's nurse might have seen *and recognised* him, in Naples at the very beginning of April.[16] Unfortunately, this information, in which the nurse speaks of Majorana "between *Palazzo Reale* and the *Galleria*, walking up from *Santa Lucia*", is at best unreliable... As a matter of fact, the buildings mentioned here are so placed that one could not deduce that a person was coming from *Via Santa Lucia* just from the fact that they were walking between the *Palazzo Reale* and the *Galleria Umberto I*, simply because *Via Santa Lucia* does not join there. This observation, which would have gone unnoticed by Majorana's relatives and those who are not familiar with the city of Naples, should instead have been obvious to the Neapolitan investigators.

In any case, the meeting between Salvatore Majorana and Senator Bocchini, organised by Giovanni Gentile, gave a new drive to the enquiries into the professor's disappearance. Indeed, on the quoted document it was clearly noted that "H.E. wishes the enquiries to be intensified", and this is what actually happened (though, from a note on the back of the document, we learn that the official in charge had "taken note and executed" only at the end of that week, on April 23). Moreover, these new enquiries by the political police led to no further positive results.

Pressure from the Ministry of Education

Although it may seem that the police were not particularly concerned about the case of the missing professor (at least, not to the extent that Ettore's family would have liked), this was surely not so for the Minister Bottai, who had granted Majorana the tenure for exceptional merit.

As soon as chancellor Salvi from the University of Naples informed the Ministry of Education in Rome of the teacher's disappearance, with the letter of March 31 quoted above, the minister himself promptly responded[17]:

> With reference to the above-mentioned letter, I request that you give me, as soon as possible, further news of Professor Ettore Maiorana.
>
> Signed Bottai.

(Footnote 15 continued)

statements, however, that we have no proof for. For example, a sailor claimed that he had recognised Ettore "on the stern of the ship, bent over the hammock netting", while two stewards said they saw him disembark in Naples, carrying a small suitcase, and one of the two recalls: "As a matter of fact, I told my colleague, given the small amount of luggage, with that passenger over there, we won't get any tips for carrying his case..." See, for instance (Ferreri and Magnano 1972).

[16]We know from a letter to his family that Majorana was a somewhat "sickly" person and had hired a nurse [see the letter MF/N2 of 22 January 1938 in Recami (1987)].

[17]This and the following letters and notes are in the MPI file.

Clearly impressed that the Minister in person had become involved, the chancellor hurried to reassure him that he was in contact with the Naples Police Commissioner, and in the letter of April 15, he wrote:

> With reference to the above-mentioned letter I regret to inform you that, up to now, we have not received any news of Professor Ettore Maiorana.
>
> This office remains in permanent contact with the local Royal Police Headquarters and I confirm that I shall communicate any positive news as soon as it arrives.
>
> The Chancellor
>
> G. Salvi.

The minister wanted to maintain a high level of attention on the case, and on April 29 he wrote back to the chancellor:

> We acknowledge the content of the above-mentioned letter regarding Professor Ettore Maiorana, and await any further news you may have of the latter.
>
> Signed Giustini.

As soon as chancellor Salvi gleaned new facts, he communicated them to the ministry (with an undated letter, which we assume would be of May 3):

> Further to my previous communications, I herewith transcribe the Police Commissioner's note no. 87965 of April 29:
>
> "As you know, Your Honour, Professor Maiorana left the Hotel Bologna of this city in March, manifesting suicidal intentions to Professor Carrelli of the Institute of Experimental Physics of this Royal University, intentions which he did not subsequently carry out.
>
> On request of his brother B.S. Salvatore, a search began and was then intensified, but up to now with negative outcome. The only thing that has come to light is that the missing person, perhaps on the 12th of this month, went to the Convent of San Pasquale in Portici to be admitted to that religious order, but as his request was not accepted, he withdrew to an unknown destination.
>
> The enquiries continue with all due diligence and, in the case of positive results, Your Honour will be informed."
>
> The Chancellor
>
> G. Salvi.

The searches were indeed intensified, as we have seen, thanks to the intervention of Senator Bocchini, and finally it seemed that a new lead was emerging, this time directly from the police investigations. We shall come to this in a moment. But here we cannot help noticing that, while the new lead regarding the convent is (of course) treated as a mere hypothesis, one thing was now sure about Majorana: he did not carry out his "suicidal intentions".

This was also the last piece of news we know of from the police headquarters. Although the enquiries were not yet over, they produced no further results.

The Majoranas, meanwhile, convinced as they were that Ettore had not committed suicide (at least as far as his mother was concerned), decided to go on in a different way, and started to publish some advertisements in newspapers. Two of these were also picked up by the Ministry of Education (and collected in Majorana's file).

The first appeared in *Il Resto del Carlino* of Bologna on July 14, perhaps through the intervention of Ettore's uncle, Quirino Majorana, professor at the Royal University of Bologna. The second, however, was published in *Il Messaggero* of Rome (where the Majoranas lived) on July 16. Here is the text of the latter:

The mysterious disappearance of Professor Majorana

The scientist may be hiding alone in pursuit of an ascetic ideal

Naples, 15 night

"Ettore, your mother and your brothers await you in despair. Please come back". This heartfelt plea appeared some months ago in the Neapolitan papers and concerned professor Ettore Majorana, who disappeared last March from Naples, as already recounted in this newspaper, shortly after beginning his lectures on theoretical physics at the Royal University of Naples, a tenure he had been granted outside the usual selection process "for high repute of particular expertise".

The very young Sicilian professor, born in Catania on August 5, 1906, was a guest at the Hotel Bologna of our city where, as we have said, he had just begun to teach. With an introverted personality, he only made friends with a single person, professor Carrelli. At the beginning of last March, Carrelli received a strange telegram from Palermo, signed by Majorana, in which the Sicilian professor expressed his decision to reach "a more complete spiritual perfection". After the beginning of a letter in which professor Majorana confirmed that he had not yet been able to reach his highest goal, there has been no news from him. From the moment of his disappearance, the most meticulous investigations have been carried out to trace him, but in vain. At the moment of his disappearance he had 3000 liras on him.

The wildest theories are circulating regarding his fate, but the one now considered most plausible is that the professor, having expressed his intention to reach high spiritual perfection, is now hiding in some isolated place, from where it may be that he will one day show up, bringing some peace of mind to his old mother, now stunned with sorrow.

Disregarding the many journalistic inaccuracies in the article, and the absence, typical of the fascist regime, of the natural hypothesis of a suicide (which would seem perfectly synonymous with a "high spiritual perfection"), let us just point out the reference, curious for a piece of journalism, to the fact that Majorana was carrying 3000 liras. As we saw earlier, such an amount of money is not plausible, although it might have been what the Majoranas had estimated. They were probably not aware of the fact that Ettore had withdrawn his salary up to February. If that is true, the 3000 liras should be added to the money corresponding to three and a half months' salary, subtracting his expenses in Naples, to get the amount he might have had with him when he disappeared.

The press reports roused the Ministry of Education, which immediately, on July 19, addressed another letter to chancellor Salvi:

News relating to Professor Maiorana's disappearance has been published in a newspaper in the last few days.

With reference to our previous correspondence, we would ask you to let us know, with the utmost urgency, what has come of the searches organised by the authorities in charge at the time.

Signed Giustini.

The chancellor's answer was not long in coming, arriving at the Ministry on July 25:

Besides the news sent to Your Honour with my note no. 901 R on May 3 of this year, no other news has arrived at this office regarding Professor Maiorana's disappearance.

I have already asked the Police Commissioner to keep me informed, but I shall in any case restate my request. I assure you that I will send on any news of further developments.

The Chancellor

G. Salvi.

This answer, which corresponded to the reality of the situation, does not seem to have satisfied the officials at the ministry, who wrote "Consult urgently", underlining the last word three times on the letter. The chancellor was probably even contacted by phone to obtain more details.

After a few days, on July 28, the chancellor wrote to the ministry again, and once again he had no significant new information to give:

With reference to my renewed appeals, the local police inform me with sheet 24 of note 138008 that their extremely active search programme to find Professor Ettore Maiorana, involving all the Police Headquarters in the Kingdom, has so far given negative results, and they assure me that they will not fail to inform me of any significant discoveries.

I confirm what I have already assured you, that I will keep you informed of any news I might receive regarding the above-mentioned affair.

The Chancellor

G. Salvi.

In the following days of August, there were other letters between the ministry and the chancellor, especially relating to the matter of the salaries collected by Majorana (which may have come to light thanks to the story in *Il Messaggero*). We have already reported on this in the previous chapter.

The Religious Lead

The appeals in the press did not bring the desired results. Therefore, on July 27, Ettore's mother—becoming more and more agitated—wrote a plea directly to the Prime Minister Mussolini (accompanied by a letter of presentation from Enrico Fermi, a member of the Academy of Italy):[18]

---

[18]Part of this letter was reproduced in the Prologue; it is reported in full in Recami (1987).

Excellency,

I turn to you, the highest arbitrator and architect of justice, to implore you to intensify as far as possible the measures adopted to locate my son, Ettore Majorana. Full professor of Theoretical Physics at the Royal University of Naples, he was appointed for exceptional merit last November.

His sudden and heart-breaking disappearance dates back four months now, and we have only one sure clue: during the last days of March or the first of April, Ettore Majorana introduced himself in a highly agitated state to the superior of the Chiesa del Gesù Nuovo in Naples and asked to be taken into experience the religious life.

Not accepted at once, for obvious reasons, he disappeared and was never heard of again. All the enquiries carried out by the ecclesiastical authorities have proved fruitless.

He was always wise and balanced, and the tragedy of his soul and nerves remains a mystery. But one thing is sure, and it is firmly attested by all his friends, his family, and myself, his own mother: no clinical or moral precedents were ever noted which might point at suicide. On the contrary, the serenity and strictness of his life and studies allow us, in fact compel us, to consider him only a victim of science.

And no one is a better witness of this than H.W. Enrico Fermi of the Academy of Italy, who was his mentor and friend, and who has addressed to Y.E. the attached letter, as an expression of the respect he has for my son.

I know that the police have worked diligently to carry out the search, but without results so far.

If I may express an opinion, it would be more profitable to seek my son in the countryside, in the home of a farmer, for example, where he would more easily have been able to avoid the thorough searches and vigilance of the police, and where he may have been able to save the little money he carried on him for a little longer.

But so far there has been no sighting, although the search bulletin named him three times. In case my son has gone abroad, I inform Y.E. that his passport (no. 194925) expires in August and will have to be renewed at some consulate.

Excellency, it is an illness caused by not unworthy studies, perhaps perfectly curable, but perhaps also destined to worsen irremediably if neglected. Your powerful intervention will doubtless decide the fate of these investigations, and the destiny of a man.

Excellency, before you who are entitled to the brightest and most generous initiatives, inspired by enlightened comprehension and always crowned by victorious success, a grieving mother kneels in confident hope.

The anguish of a mother over her son's fate is undisguised throughout the letter, but we can nevertheless still extract some useful information. Listing the several possible leads, she neglects no plausible hypothesis, and yet she discards the suicide hypothesis out of hand. As a matter of fact, when the Naples Police Commissioner showed Ettore's brother the missing person's file, where the *Duce* himself had written "I want him to be found",[19] the Commissioner noticed that "a living man cannot be found, a dead one can". We see, then, that the letter opens with the

---

[19]According to oral statements by Ettore's relatives; see Recami (1987).

sighting at the *Chiesa del Gesù Nuovo* in Naples, expressing what relatives believed most reasonable at the time.

As already mentioned in previous pages, the "religious" lead, according to which Ettore may have sought shelter in a convent or something such place, was followed up by his relatives right from the beginning of April, well before Strazzeri's suggestion to Salvatore Majorana. As Maria, Ettore's sister, recalled:[20]

> Searches have been carried out in convents. Mom was talking about that with the parish priest, who has written to many convents. Later on, after a certain time, we also sent a petition to Pope Pius XII.

Only two sightings seemed genuinely "positive", however: the one on April 12 from the Convent of S. Pasquale in Portici (on the outskirts of Naples), appearing in the Police Commissioner's note to chancellor Salvi of April 9, already quoted, and the earlier one, which only came to light later, from the Jesuit *Chiesa del Gesù Nuovo* in Naples, mentioned in the letter from Ettore's mother to Mussolini. Although the former information came from the Naples Police Headquarters, nobody ever paid any further attention to that, and it is not even mentioned in the letter to Mussolini, a sign of its presumed lack of credibility. As regards the latter, on the other hand, it may not even refer to the days *after* Majorana's disappearance. Edoardo Amaldi, who got first-hand information directly from the relatives of his friend Ettore, actually gave some further details:

> No clue was ever found: we only learned that, some days before Ettore Majorana's departure for Palermo, an agitated young man, whose physical and psychological features were – according to his relatives – similar to Ettore's, arrived at the Chiesa del Gesù Nuovo, in Naples near the Hotel Bologna where he was living. Besides, Father De Francesco, a former Jesuit Provincial Superior who had welcomed the young man, thought he recognized him in the picture of Ettore shown to him by his relatives. The young man asked Father De Francesco to "try the religious life", an expression which, according to the brothers, has to be interpreted as "undertaking spiritual exercises". In fact, they did not believe he wanted to manifest a religious calling by this remark, but simply the desire to withdraw in meditation. To the answer that he could remain there, but only for a short period of time – because for a permanent solution the Congregation would require him to enter the Novitiate – the young man answered: "Thank you, sorry", and went away (Amaldi 1968).

We do not know when the new sighting came out, but it is curious that, despite chancellor Salvi's pressures, what the Majoranas felt was a "sure clue" was not to be found at the end of July in the note from the Naples Police Headquarters. Clearly, the information checked by the relatives (and not by the police) was either considered invalid, or they had already decided that it was not directly relevant to the disappearance.

In the case of Majorana's disappearance, this religious lead looks quite peculiar because, in Strazzeri's words, it is about a "not very religious person", as confirmed by the brothers themselves, who did not believe he "wanted [...] to manifest a religious calling". Nevertheless, such a lead, as we have seen, had already come to

---

[20]Ponz de Leon, television programme, *loc. cit.*

light a few days after Ettore's disappearance. Given that the two "positive" sightings were eventually judged unreliable, it is interesting to try to understand *why* Ettore's relatives followed this path, which was not the most natural one, so readily and so thoroughly.

In this regard, the brothers' interpretation, quoted by Amaldi, is quite intriguing: Ettore wanted to "undertake spiritual exercises" (the way Jesuits do). As a matter of fact, a curious episode has recently come to light, which may date back to the 1950s. It is told by one of Carrelli's former collaborators, Elio Tartaglione.[21] One day, after giving a lesson at the university, the two of them met to visit the church and convent of *San Gregorio Armeno* in Naples, not far from the Institute of Physics (inside, there is a beautiful cloister, particularly significant from an artistic point of view, and which was restored at around this time, after the tragic bombings of the World War). While showing Tartaglione some windows of the convent, Carrelli mentioned that Ettore Majorana was once hosted there for spiritual exercises... When asked for further information, Carrelli changed the subject at once. This information was also known to Majorana's former co-ed, Gilda Senatore,[22] who recounts that the convent allegedly took Majorana in for "exercises" for about two months[23] (Her source was, however, Tartaglione himself, one of her husband Cennamo's friends and colleagues).

Carrelli was rather talkative by nature, but it seems reasonable to believe, from the professionalism of his scientific and personal life, that the information quoted above is well-founded (although it may well be that this is an interpolation, encouraged perhaps by the enquiries made by the Majorana family). If this is the case, it is easy to find a simple explanation for the importance assumed by the religious lead and the Majoranas' commitment to it: Carrelli could conceivably have reported all the information he had to Salvatore Majorana, and this would have encouraged the searches in the convents. Even if this circumstance is confirmed, we cannot help noting that *Ettore himself* must have let his Neapolitan friend Carrelli in on the matter (perhaps without exaggerating certain things, such as the two-month duration). Now, even leaving aside, as did the family, the improbability of Ettore's going through a "spiritual crisis", it is a fact that in the relevant period of time

---

[21]See, for instance, Elio Tartaglione's letter to Bruno Preziosi, read by the latter at the XXVI National Congress of the Italian Society of the Historians of Physics and Astronomy, Naples, June 3, 2004.

[22]Esposito's interview with G. Senatore, *loc. cit.*

[23]Actually, perhaps through simple carelessness, Senatore said that Carrelli and Tartaglione passed by the *Chiesa di S. Aspreno*, rather than *S. Gregorio Armeno*, while walking to the institute. Though this version is plausible too, we shall opt for Tartaglione's, as it comes to us directly. In any case, curiously enough, the *Chiesa di S. Aspreno* in Naples, unlike *San Gregorio Armeno*, is near the already mentioned *Piazza Bovio*, on the street from the *Albergo Bologna* to the Institute of Physics.

(and even now to some extent) the convent of *S. Gregorio Armeno* was a cloistered convent of nuns, where men would certainly not have been admitted![24]

So this question, too, *would seem* destined to remain unsolved...

A Fictional Interlude

In the Majorana file opened by the fascist political police, after the meeting between Salvatore Majorana and Senator Bocchini, there is one last page where we may read the following—besides the above-mentioned material (i.e., the meeting request, Giovanni Gentile's letter, and the note of April 18):[25]

Rome, August 6, 1938

> Still regarding some movement against Italian interests, rumour has it in some quarters that the disappearance of Majorana, a man of great value in the field of physics and especially radio, the only one who could carry on Marconi's studies in the interest of national defence, may be the victim of some obscure plot to get rid of him.

In conclusion, it occurred to some diligent fascist to link the disappearance of the physicist Majorana to some spy story, although being a pure theorist, he could not have and would not have wanted to "carry on Marconi's studies". The political police, who would not neglect any such "tip-off", decided to keep this note in the missing person's personal file; perhaps Mussolini himself was informed about that (we cannot help noticing that this note came a few days after the letter from Ettore's mother to the *Duce*).

After a few years, in 1944, when leading the Republic of Salò, Mussolini was informed that there was an Italian in the group of scientists who were working in Germany on Hitler's secret weapon, capable of completely transforming the fate of the war. It seems[26] that the missing scientist from 1938 came to the *Duce*'s mind on this occasion, and he appointed Filippo Anfuso, his ambassador in Berlin, to carry out enquiries along these lines to obtain some kind of confirmation. However, these were not completed because Nazi Germany then collapsed, and there remains no trace of the letters between Mussolini and Anfuso.

After the Soviets captured the scientists in one of the German teams working on the atomic bomb, the one directed by von Ardenne (the other, intercepted by the allied secret mission Alsos, was directed by Heisenberg), it seems that the *Gazette*

---

[24]Tartaglione may have made a mistake here, by not realising that Carrelli was perhaps pointing at the windows of the all-male *San Lorenzo* convent, right in front of *San Gregorio Armeno*. However, this is of no help in solving the problem, because the holy order operating from the convent of S. Lorenzo does not generally practise "spiritual exercises".

[25]Note inside the Political Police File.

[26]This and the following pieces of information can be found, for example, in (Poggio 1972), or in the book (Castellani 1974), but have no confirmation. We are reporting them merely out of historical curiosity, since they were occasionally discussed between the 1960s and the 1970s.

de Lausanne revealed, we do not know on what grounds, that the Soviet government had done its best to come into the possession of some of "Majorana's notebooks".[27]

These two tempting pieces of news, coming many years after the Sicilian physicist's disappearance (and during the cold war), were the basis of several tales according to which Majorana was involved in the development of the atomic bomb. These stories were encouraged by something Ettore is supposed to have said to his friend Carrelli on several occasions, as divulged by Carrelli himself: "Physics has taken a bad turn, we have all taken a bad turn".[28] As far as we are concerned, we here endorse the opinion of Ettore's friends and relatives, as quoted by Amaldi:

> Only thirty or so years later did someone who had never known him, or had known him only very superficially, imagine a kidnapping or an escape relating somehow to a supposed case of atomic espionage. But for those who lived in the nuclear physics community of the time and who had met Ettore Majorana, such a hypothesis not only has no basis, but it is nonsense, on both a historical and a human level (Amaldi 1968).

Although at the time of Ettore's disappearance, the rush to exploit the energy of the atomic nucleus was in the air (advocated mainly by Leo Szilard), it was not until the end of 1938 that Hahn and Strassmann's crucial experiment on nuclear fission finally gave some results (published at the beginning of 1939)—and in itself this only opened up the long road to eventually constructing the atomic bomb. Majorana, despite being a "first-class theoretical physicist", surely did not have supernatural powers of prediction regarding future events.

For the sake of truth, we should not hide the fact that these two pieces of news came to light *without* any knowledge of the note made by the political police on 6 August 1938.[29] But, beyond the coincidence and the improbability of the three pieces of information, this should be considered only as an indication of the wonder inspired by the disappearance of a scientist like Majorana, even in the most practically-minded observers.

An intriguingly new line of supposition, with a hint of spy story in it, comes from the fact that also the U.S. Central Intelligence Agency was interested in Majorana during the 1950s (and, possibly, later). Indeed, from the public domain documents published by C.I.A. on their website, we learn that an entry was created ("Atoms Oriented in a Variable Magnetic Field") in December 1951 about Majorana's paper P6, while another ("Symmetrical Theory of the Electron and the Position") appeared in August 1957 about paper P9. Probably, paper P6 was

---

[27]This news, reported in the quoted articles, and which looks highly circumstantial, has also turned out to be inaccurate. In 1946, many articles were published in the *Gazette de Lausanne* that related in some way to work on atomic weapons (the hydrogen bomb). Recall that in July 1946 the first such weapon was exploded over the well-known Bikini atoll. But it is easy to check that none of the articles in this newspaper was about Majorana and his disappearance.

[28]See (Poggio 1972).

[29]Otherwise, we would have to assume that the content of this note was disclosed at least to Majorana's relatives, and that these in turn would have disclosed it to others. Of course, this seems most unlikely.

interesting to C.I.A. because of nuclear magnetic resonance (Bloch and Rabi 1945), while paper P9 (disregarding the curious mistake: "position" instead of "positron") because of the appearance of Pontecorvo's papers on neutrino oscillations in Soviet Union (Pontecorvo 1958b). However, whatever the motivation for the creation of these entries, it certainly does not hide anything mysterious: other entries appear that are related to papers by Quirino Majorana—Ettore's uncle—who surely was not a person of interest for C.I.A.. Something magic, if not mysterious, lies in the name of Majorana…

Chapter 10
The Last Chapter

News from Sicily

After the note by the political police on August 6, essentially no other relevant information came to light on the case of the missing professor. But we cannot leave the "case" without some reference to the enquiries carried out in Palermo, where Ettore should have arrived on the morning of March 26.

His friend Emilio Segrè, who was then professor at the University of Palermo and, as the reader may remember, had announced the new selection for the tenure in theoretical physics, carried out the enquiries there with the Majoranas:

> We knew nothing of Ettore. His brother, the engineer Luciano, who had been a school-friend of mine, rushed to Palermo and together we tried to find Ettore with the help of the police. We found only that he had stayed at the Albergo Sole, as was already clear in the letter (Segrè 1993).

This is Segrè's only mention (together with the statements by Majorana's relatives) of the searches carried out in Sicily "with the help of the police", who strangely enough, as far as we know, did not issue any official note to be included in the Majorana file.

Two of Ettore's friends lived in Sicily, Segrè and Wick, who was granted the theoretical physics tenure and settled in Palermo in January 1938. It seems the latter did not participate actively in the search. However, there were also Ettore's cousins and other members of the large Majorana family, from Catania. From one of these, Claudio Majorana, a member of the Italian Parliament and uncle Dante's son, we have the following testimony[1]:

> It turns out, as already known, that Ettore went to Palermo to meet Emilio Segrè. He had informed us about his journey to Sicily in a letter, asking us to excuse him for not stopping by in Catania, because of the short time available to him. He did not tell us anything about

---

[1]216 See the interview with Claudio Majorana in (Randazzo 1972).

© Springer International Publishing AG 2017
S. Esposito, *Ettore Majorana*, Springer Biographies,
DOI 10.1007/978-3-319-54319-2_10

the reasons for his visit to the famous scientist. There was no meeting in Palermo, because Segrè was away, so my cousin returned to Naples.

What cousin Claudio is reporting here sounds a bit strange, as we have no other trace of this information. Despite some inaccuracies in the text of the honourable Majorana's interview (on things referred to by others), what is stated above (which only the cousins from Catania would have been aware of) is very well circumstantiated, and we are thus inclined to accept its overall content. If we assume good faith, some distant recollections may have confused the journey to Palermo with other journeys made by Ettore. The same cousin Claudio remembered, on the other hand, that "from 1936 to his disappearance [Ettore] returned to Catania only a few times and always for short periods of rest". Anyway, the reason for Ettore's visit to his friend and colleague Segrè, who had moved from Rome to Palermo in 1936, remains a mystery.

Ettore's alleged accommodation in Palermo, the Grand Hotel Sole, is puzzling, too. Located in *Via Vittorio Emanuele*, near the cathedral, it is a long way from both the port and the Institute of Physics and, despite its name, it would not really appear to be as "suitable" for a member of the Majorana family as the nearby Hotel Centrale in the same street. In this area, even closer to the port, there were many other hotels,[2] so it is difficult to understand the reasons for such a choice. The only thing we might note is that the hotel run by Vincenzo Sole stands near Palermo's Central Post Office, from where Majorana sent the letter to Carrelli, quoted earlier. This letter is also peculiar, because it is actually the only one in Majorana's correspondence written on the hotel letterhead. In the letters to his family and friends (and, almost without exception, all those related to his scientific and teaching activity), Ettore did not usually use headed paper, even though he always noted down the place he was writing from. Of course, this may be purely by *chance*...

As we have seen, the only witness to confirm a possible return of the missing person from Palermo to Naples was professor Vittorio Strazzeri of the University of Palermo. A widower with three children, he was born in Gela, near Caltanissetta, in 1874, so in 1938 he was 64. He became (adjunct) professor of analytic and projective geometry, and also descriptive geometry with drawing, only in 1935, coming second in the selection organised by the University of Messina, after teaching in high schools for many years (when he won the selection he was teaching in the Art High School in Palermo). The committee for this selection was made up of E. Ciani, chairman, U. Amaldi, G. Scorza, G. Marletta, and E. Bortoletti, secretary. Here is their summary of his professional activities[3]:

Vittorio Strazzeri – Graduate in mathematics (Palermo, 1898); teacher of mathematics (since 1898) then mathematics and physics (since 1923) in Royal Middle Schools. Voluntary assistant for advanced analysis (1912-13), then analytic and projective geometry

---

[2]For this purpose, for instance, the reader may consult the official telephone directory for Sicily of 1938.

[3]See *Bollettino del Ministero dell'Educazione Nazionale* (Part II: Administrative Acts), year 63, vol. I, p. 210.

(1914-19), then astronomy (1923-25) at the Royal University of Palermo. Since 1917, guest lecturer on analytic and projective geometry. Appointed in 1919, and from 1928 on, for the course on descriptive geometry in Palermo; from 1923 to 1927 appointed for advanced geometry (ibid.)

He is presenting 29 publications, which can be classified into four groups [...].

Strazzeri's industriousness, his mastery of the algorithms he uses, and particularly his ability to systematically rewrite theories are noteworthy; also commendable for someone who has spent 37 years teaching middle school are his constant interest in scientific research and his ability to take into account modern developments. Not all of Strazzeri's work is relevant; but in his preferred field of geometry, he has made considerable contributions, and has proven himself capable of good taste and geometrical imagination.

He is well known as a gifted and effective teacher [...].

Strazzeri's activity as a guest lecturer in the following three years, before being granted tenure, was summed up as follows by the committee set up for the purpose (whose members were F. Severi, G. Scorza, and N. Spampinato):[4]

[...] his scientific activity has been somewhat scarce; but the topics he has studied always have that elegant quality already encountered in the candidate's previous work on the occasion of the selection which brought him to his present teaching position.

Hence, the committee, considering the significant seniority of the candidate, who was for many years a talented middle school teacher, proposes that he should be granted tenure.

Oddly, Segrè and Strazzeri became professors at the University of Palermo in the same year (December 1, 1935), the former at the Institute of Physics and the latter at the School of Mathematics, whose buildings were both in *Via Archirafi*, one next to the other. And it is equally odd that, for a while, the two men lived very close to each other, Segrè at *50 Via A. Borrelli*, near the luxurious Hotel Excelsior, and Strazzeri at *15 Via Pepe*,[5] in a residential area quite far from the city centre. They probably knew each other well, and it might have been thanks to Segrè that the Majoranas acquired Strazzeri's testimony.

Even though we do not have first-hand evidence, it is plausible that Strazzeri did indeed travel on the Palermo-Naples mail boat on March 26, 1938. As a matter of fact, the professor from Palermo travelled a great deal for study reasons. For instance, in April 1937, he attended the first Conference of the Italian Mathematical Union in Firenze,[6] and was also invited to edit a volume (the VI, on the congruence of lines and spheres and their deformation) of the national edition of work of the famous Italian mathematician Luigi Bianchi, who introduced into Italy the theory of groups which so fascinated Majorana and his friend Gentile for its universality. In addition, at least from 1936, Strazzeri used to go to Germany every year, during the

---

[4]See the *Bollettino del Ministero dell'Educazione Nazionale* (Part II: Administrative Acts) year 66 (1939), vol. I.

[5]From the Annual Academic Report of the Royal University of Palermo for the year 1936 (and following).

[6]See *Atti del primo Congresso dell'Unione Matematica Italiana tenuto in Firenze nei giorni 1-2-3 aprile 1937*, Zanichelli, Bologna, 1938.

summer holidays, ostensibly for "study purposes", as we can see from his authorisation requests to the Ministry.[7] However, it is quite possible that the main reason
for these trips to Germany was that one of his daughters, Maria Cristina, was
married to a German doctor, Christian Scharfhillig, so he may just have been going
there to visit his relatives. It is interesting to note that the main railway line from
Italy to the German town where they lived, Frauenburg in eastern Prussia (today
Frombork, in Poland) passed through Leipzig. Who knows whether Strazzeri may
have done this trip in 1933, when Majorana was working in Leipzig...

A Curious Observation, but not Completely Crazy

"The young man, acutely misanthropic and not in good health, may have retired to
some destination in Palermo, [...] or to some private clinic. [...] It is unlikely that
he has gone anywhere he may be recognized". As we have seen, from this starting
point, the police started searching for the missing person, and there is no reason to
doubt that the detectives really did carry out inquiries in hotels, hospitals, and
clinics, both in Naples and in Sicily. However, from the documents and witness
statements we have, we do not know whether they made inquiries inside the Naples
mental hospital or, as it was called at the time, the *Ospedale Psichiatrico
Provinciale "Leonardo Bianchi"*.

This apparently bizarre observation may seem more relevant if we note that such
a place might be considered more appropriate than a convent for someone "not in
good health"—looking for a quiet life away from prying eyes. Moreover, as an
example, in May 1938 there was a famous patient at "Leonardo Bianchi", who had
known Majorana for a few months: the mathematician Renato Caccioppoli (Toma
2004). His "reclusion" was organised by his family (at least for the whole of the
period when Hitler was in Naples): they wanted to keep the antifascist Caccioppoli
safe from the political police, and an asylum was the perfect place to find shelter.
Someone even took the mathematician's "madness" seriously, and it is curious that,
still a year later, the chancellor of the University of Naples, when he wrote to the
Minister of Education on 25 April 1939, drew a parallel between the mathematician
Caccioppoli and the physicist Majorana.

> Professor Caccioppoli is to be considered as mentally unbalanced (something we hope will
> be temporary) and therefore he must be considered as a neurotic, who cannot completely
> control himself and cannot adequately perceive and evaluate the various events and
> moments of social life, something often seen in individuals whose intelligence is one-sided
> and who, completely absorbed in studying subjects requiring intellectual focus and par-
> ticular self-denial, are almost completely estranged from other aspects of life. Hereto, we
> cannot help thinking of the recent case of Professor Maiorana [sic!]. And it is neither

---

[7]This information is in Vittorio Strazzeri's file at the Ministry of Education, *Direzione Generale
Istruzione Superiore, Fascicoli Personale Insegnante e Amministrativo*, kept at the Central State
Archive in Rome.

inappropriate nor superfluous to recall that we are talking about very young men who obtained tenure at an age when the others are still in their formative years; they found themselves in a position of responsibility, such as that of university professor, and they are totally unprepared to cope with the studies and the needs of an environment they were totally unaware of during the strict and absorbing studies which led them to that very tenure (Bartocci et al. 2007).

But Caccioppoli's and Majorana's cases were certainly quite different. For Majorana, discretion would not have been guaranteed, as it would have in a convent, unless there was an "accomplice"—and one alone—inside the mental hospital. Well, it is clearly most unlikely that, in the short time he stayed in Naples, Majorana would have met someone working in that hospital, who might have helped him to find refuge there. But perhaps the contrary is less unlikely.

As a matter of fact, as we have seen, Carrelli was a very effective "propagator of news". A friend of Giovanni Pisapia, Gilda Senatore's uncle, he was probably the one who told Majorana about his student's South American origins. So it is also likely that the director Carrelli had introduced Majorana to his colleague from the *Accademia Pontaniana* in Naples, the director of the *Ospedale Psichiatrico* Carrelli knew so well (Bonolis 2008). Then again, he was also the father of another of Majorana's students: the director of the *Ospedale Psichiatrico* was *Michele Sciuti*, son of the late Sebastiano. The coincidence becomes even greater when we realise that Michele Sciuti was not from Naples like his son, but came from an important family in Catania. Indeed, their family home was at *5 Via Paternò*, not far from *251 Via Etnea*, where Ettore's family had always lived (although, of course, this was just a coincidence).

But now the coincidences seem to be growing more and more numerous, and it is worth focusing on this curious matter for a while.

Michele (or more properly, Michelangelo) Sciuti was born in Catania on April 29, 1875, into a well-known family (just like the Majoranas, in fact), and he could boast several famous relatives, such as the painter Giuseppe Sciuti, among others (Calvesi and Corsi 1989). He came to Naples to study medicine, graduated at a very young age, and soon entered that "group of scholars, thirsty for knowledge, gathered around *Maestro* Leonardo Bianchi who, in his university tenure, taught and carried out research on the nervous system and mental pathology in the large rooms and laboratories of the ancient and glorious Institute of *San Francesco di Sales*" (Vizioli 1948). From 1901, Sciuti was one of the doctors of the *Manicomio Provinciale* (Provincial Mental Hospital) in Naples, first as junior doctor, then as general practitioner, head physician, and finally, from 1925, director (he remained in charge until his retirement in 1942, when he became Director Emeritus).

Sciuti's work in organizing the Neapolitan Psychiatric Institute, the intellect and heart he put into this endeavour, and his total devotion to the task are attested by the excellence and distinction that his hospital achieved, thanks to him, both in the field of caring for the infirm and in the field of scientific activity. He was, among other things, a committed supporter of occupational therapy; he set up important manufacturing facilities for his patients, including weaving rooms, a tile and concrete factory, and a perfect and well organized printing press (Vizioli 1948).

Beyond the likely rhetoric of a commemoration, there can be no doubt, as one can see from the available documents,[8] that Michele Sciuti had a strong impression both on his patients and their relatives and on the scientific community for his important work in histology, pathology, and neuro-psychiatric practice, known also abroad. His broad knowledge, which went well outside the boundaries of his discipline, is clearly attested by his son Sebastiano, when he remembers his decision not to follow in the safety of his father's footsteps, but to enrol in physics – and what is more, to join Fermi's group in Rome:

> When I enrolled at university – I was 17 – I said to my father: "Dad, send me to Rome". And he said: "Son! – in an old-fashioned Sicilian way – "Son, you're so young, if you go to Rome what will become of you?" So I stayed there. [...] This attraction [to physics] was also due to the fact that there was this very nice educational review, whose name I can't remember right now, my father had a subscription to; there were a lot of interesting topics, and there I found a paper by Corbino which struck me (Bonolis 2008).

More than the man of learning, what matters here is the character's personality. In this respect, it is interesting to note that, in a fairly dry text such as an obituary (Vizioli 1948), there is room to mention that he was "a profoundly and truly good man, with a big heart", "who kept his warm smile, and who always had a good and kind word for everybody". Though we have no direct evidence for this, it is very likely that, as happened in Trieste and other mental hospitals in Italy, Michele Sciuti hid some Jews fleeing from Nazi and Fascist persecution in "his" hospital, perhaps even from 1938.[9]

Given this general context, the idea that the police ought even to have looked for Majorana at the psychiatric hospital in Naples does not seem so absurd after all. Although no result has come to light from possible inquiries between 1938–9, this should not prevent further investigations from being carried out *now* in the archives of the hospital.

Anyway, here as with regard to other possible leads, it seems as if a sort of "Majorana curse" has cast its spell on these inquiries. As a matter of fact, a careful search at the former *Ospedale Psichiatrico "Leonardo Bianchi"* has only provided information about Michele Sciuti. Officially, no "Majorana" (or similar surnames) was ever a patient at the hospital, and nor did anyone (under a false name) with his features and age check in during the first six months of 1938. Of course, this negative finding does not necessarily imply any conclusive result, so further inquiries should be carried out. But as we were saying, *"chance"* has intervened

---

[8]In Michele Sciuti's personal file, kept in the archive of the former Psychiatric Hospital "Leonardo Bianchi" of Naples, now *Polo Archivistico Sanitario* of the *Regione Campania*, there are several pieces of evidence which illustrate the director's welcoming, but reserved character.

[9]This intuition, together with the possible involvement of Michele Sciuti in Majorana's disappearance, was suggested by Anna Sicolo, director of the *Polo Archivistico Sanitario* of *Regione Campania*. Our warm thanks go to her.

again and the archives of the run-down hospital have since been closed, having been rendered impracticable, so it has not been possible to complete the search.[10]

Therefore, this leaves us wondering whether that "Sciuti" Majorana referred to in his last letter to Carrelli is the student-son or the director-father; in other words, whether it was just an affectionate greeting Professor Majorana was giving to his only diligent student, or a subtle suggestion (even a subtle mockery) from an ironical Ettore.

Nearing the End of the Enquiries in the Majorana Case

The official police enquiries in Naples, Palermo, and more generally, all over the Kingdom of Italy, went on for quite a long time, in fact, about a year, without reaching any result. On April 4, 1939, the director of the Customs and Transport Police, based at the Home Office in Rome, wrote to the General and Classified Affairs Division dealing with the Majorana case:[11]

> With reference to note no. 300/45151 of March 26, 1939, please let us know whether the following alert, which is still recorded in the supplementary border sections of circular 442/10638 of March 31, 1938, should be erased or moved to the print section.
>
> Director Head of Division
>
> Saporiti

Circular 442 contains the telegram sent to all the police commissioners of the kingdom, already quoted above, and the request clearly aimed to determine whether they should or should not maintain the alert for Majorana. The answer of April 22, 1939, was laconic: *remove.*

Sometime previously the bureaucratic machine of the Ministry of Education had been set in motion.

The meeting of the Faculty of Sciences at the University of Naples, of October 17, 1938, also attended by Carrelli, issued a statement about the tenure of theoretical physics, which was free now that Majorana had gone:[12]

> On the suggestion of the Council Chairman, given the continued absence of, and lack of news concerning Professor Maiorana, tenure of theoretical physics, it is approved that the replacement teaching for this important discipline should be entrusted to Professor Antonio Carrelli at the expense of the State [...].

---

[10]Dr. Anna Sicolo's intuition, referred to in the previous note, was inspired by the fact that, in the archives of the mental hospital, she found a tag among those of the other patients which, unlike the others, carried only the word "Majorana" on it and nothing else, without reference to any medical chart. Unfortunately, because of what happened later on, it has not been possible to find the tag again.

[11]Document in the PS file; see Recami (1987).

[12]Document in the MPI file.

The Secretary The Chairman

Signed Imbò Signed Pierantoni

This proposal was sent from the chancellor to the ministry on the following October 26,[13] and on December 6, the Director General prepared a "note for H. E. the Minister":[14]

Excellency,

During the last few days of March of this year, as is common knowledge, Professor Ettore Maiorana, tenure of theoretical physics at the Faculty of Mathematical, Physical, and Natural Sciences of the Royal University of Naples, abandoned his teaching post and the city of Naples, expressing suicidal thoughts, which he did not immediately carry out, to Professor Antonio Carrelli of the Institute of Experimental Physics at the above-mentioned university.

Upon his brother Dr Salvatore's request, searches were made and subsequently intensified, but with negative results.

Given Professor Maiorana's continued absence, the Board of Directors of the Royal University of Naples proposes that the replacement teaching of theoretical physics should be entrusted, at the expense of the State, to Professor Antonio Carrelli.

I await your instructions, Excellency, regarding this matter.

Respectfully Yours,

The Director General

The note that was actually given to the Minister Bottai contained a slightly different version of the original one quoted above, more focused on the administrative order which should have declared Majorana as resigning than on Carrelli replacing him. Anyway, from the first version, we clearly understand the ministry's opinion regarding their professor's disappearance: they interpreted it as a suicide which, however, "he did not *immediately* carry out" after his disappearance.

In any case, a week later the minister issued the resignation order without hesitation:[15]

The Minister Secretary of State for Education

---

[13]This letter, also in the said file, was approved by a diligent official who noted that Carrelli was "of Italian race" and had been a registered member of the National Fascist Party since November 6, 1932 (on the occasion, therefore, of his selection for the University of Catania). Clearly, the race laws, only just promulgated (1937) in Italy, were beginning to take effect. Curiously, we can also see that a month later, on November 26, 1937, the chancellor sent the ministry the minutes of the oath Majorana had taken on January 19, 1938; the whole bureaucratic matter about his appointment, as one may remember, was dealt with much earlier, between the end of January and the beginning of February.

[14]Note kept in the MPI file.

[15]The decree is dated December 6, 1938, but the original note by the (secretary of the) minister is from December 19. The decree is in the MPI file, and is reported in Recami (1987).

According to clause 109 of the Consolidated Act of the higher education laws approved by Royal decree on August 31, 1933, no. 1592; and clause 46 of the Royal decree of December 30, 1933, no. 2960;

Considering that Professor Ettore Maiorana, tenure of Theoretical Physics at the Royal University of Naples, has left office, without valid reasons, for a period longer than ten days, starting from March 25, 1938;

Considering that, despite the searches, no news of the said professor has been obtained;

Decrees

With effect from March 25, 1938, that Professor Ettore Majorana, tenure of Theoretical Physics at the Royal University of Naples, is declared to have resigned from his office.

This decree will be sent to the Court of Auditors for registration.

Rome, December 6, 1938

The Minister

Signed Bottai

The ministerial decision was communicated to the chancellor of the university in Naples the following December 20, whence the university no longer had a professor of theoretical physics.

A Lead from South America

While the attention on the "Majorana case" had essentially faded away by the first months of 1939, the interest in the physicist, and in the missing man, has lasted much longer, right up until today. Revived every now and then by one circumstance or another, it has produced new alleged sightings or theories, some of which we have described in the previous pages. Hence, in the first half of the 1970s, after the publication of the first biography by Amaldi (1966) and a successful television programme,[16] a tremendous interest in the missing physicist was aroused once again in Italy, and many articles, interviews, and testimonies appeared in the media. Some of this material, but only that of direct witnesses, has already been used here.

At the end of 1978, some new "revelations" appeared in the press, which would go on to form the basis of the "Argentinian lead". The documents relating to this are in part contradictory, so it would not be useful or wise to rely too much on them. However, since some important scientific personalities, such as Tullio Regge, Yuval Neeman, and John Archibald Wheeler, have shown interest in this idea, we shall try to consider here only what appears reasonably certain among the many pieces of information available to us.[17]

---

[16]L. Castellani, *Ipotesi sulla scomparsa di un fisico atomico*, television programme broadcast on Italian television in April 1972.

[17]For careful review of this information see Recami (1987).

In an interview in October 1978, Carlos Rivera, then the 52-year-old director of the Institute of Physics at the Catholic University of Santiago de Chile, said this:[18]

In 1950 I went to Buenos Aires with my wife and stayed at Mrs. Frances Talbert's boarding house. This lady had a son, Tullio Magliotti, who had graduated in electrical engineering. The day before my departure for Germany, I was in my room writing on some sheets of paper. I was writing about Majorana's statistical laws, and his name appeared in large letters on one of my sheets. As soon as Mrs. Talbert saw the papers she cried out: "Majorana? But that's the name of a very famous Italian physicist who happens to be a close friend of my son's. As a matter of fact, they see each other very often. My son told me that he is not involved in physics any more, but in engineering". The conversation went on for a while and the lady said: "Majorana told my son he had left Italy because Enrico Fermi didn't like him. Actually he said more: he didn't even want to hear Fermi's name anymore". According to the engineer Magliotti, this aversion came in part from the fact that Fermi was "a difficult chap", and in part from the fact that he had played an important role in the development of the atomic bomb. The conversation was interrupted by a phone call from her son. Perhaps he did not like the fact that his mother had spoken to me about Majorana and that I wanted to meet him. In fact, Mrs. Talbert did not come back to resume our conversation, and since I was supposed to embark for Germany the following day, I did not have the opportunity to meet her son or talk to her again. Four years later, in 1954, I returned to Buenos Aires and went to visit Mrs. Talbert. But the door of her house was sealed and there was nobody inside. I asked for information among the neighbours; they told me that mother and son had suddenly and mysteriously disappeared. Mrs. Talbert was openly anti-Peronist, so I would guess that they have both been eliminated by the Peronist police. I also went to check the engineer register, but Magliotti's name did not appear there. Killed? Fled somewhere outside Argentina? I just don't know [...].

In 1960 I returned to Buenos Aires for the third time. I took a room at the Continental Hotel. This is where the paper napkin episode occurred. While I was writing formulas on one of these napkins at my table, the waiter said: "I know another man with this habit of writing formulas on paper napkins, just like you. It's a customer who comes every now and then for coffee and his name is Ettore Majorana. This man was an important physicist who ran away from Italy many years ago." This latter episode, though less important than the former, convinced me that Majorana must be in Argentina. The waiter did not know where it might be possible to find him.

The news came out essentially thanks to the interest of the Israeli theoretical physicist Yuval Neeman. In his own words:[19]

My interest in Majorana was essentially reawakened by certain conversations with the late Racah [...].

I am responsible for the revival of the "Argentinian version". At first it was Wheeler who came to know about it from Meinhardt (a Jewish Chilean physicist, who in the meanwhile had emigrated to Israel) in Varenna in 1977. [...] So I looked for Meinhardt in Israel and arranged a meeting with Tullio [Regge] when Tullio visited me in May 1978. Tullio went to Chile from Tel Aviv and collected details from Carlos Rivera.

---

[18]See G. Gullace's article in the Italian magazine *Oggi*, October 14, 1978.
[19]Y. Neeman's letter to E. Recami of October 20, 1980 in Recami (1987).

The information from the interview was in fact checked by another important theoretical physicist, Tullio Regge,[20] and then confirmed in person to Recami[21] and Ettore's sister, Maria Majorana, though Rivera was more cautious with her[22]:

Mrs. Talbert was advanced in age (she was a friend of my mother's, who had met her during a trip from France to Argentina); she was completely convinced that the man was Ettore Majorana, for whom she cared a lot. Though she had never met him before, when he was working in Italy with Fermi, she was confident that everything sounded certain to her.

Recently, Carlos Rivera's testimony has been confirmed by his wife Violeta (after her husband's death), although some of the details differ slightly:[23]

While [Carlos Rivera] was reading his physics book, Mrs. Talbert asked him: "What are you concentrating so hard on?" He answered: "I'm studying Majorana's forces." The lady was very surprised and said: "Majorana?" My husband answered: "Ettore Majorana". She added: "But, Ettore Majorana is a friend of my son's." Then Carlos asked: "But do you know Majorana? Where is he?" Mrs. Talbert said: "Yes, but speak softly, don't speak too loud, because this is not a safe place. For political reasons, we can't talk about Majorana." This is what Mrs. Talbert told my husband. And she said that her son had disappeared; she did not know where he was, and was very scared. All this was happening in 1950.

And again:

It was February 1961 when we flew to Buenos Aires. Carlos and I had recently got married, and I have nice memories of the city.

I remember one day having lunch at the Continental Hotel. We were waiting at the table for our orders to be taken, and in the meanwhile we were writing some equations, as we were talking about physics. My husband was writing the equations on the napkin we had there. At some point the maître d'hôtel arrived to take our orders. When he realised that we were writing equations, he was surprised and said: "Are you physicists? You are writing physics equations…" I was immediately attracted by the waiter's broad knowledge, his Italian accent, and his interest in physics. He asked Carlos what he was doing in Argentina, and Carlos explained that he was carrying out a mission on behalf of Valdivia Universidad Austral. Then the waiter said: "If you are a physicist, you'll certainly know this person." And from his wallet he took out a newspaper cutting where there was a small photograph and something written. Carlos saw the picture and said: "But this is Ettore Majorana! Of course I know him, he is a very famous physicist!" The waiter told Carlos that over the past few years the gentleman in the photo had dined many times at that restaurant, and that he, too, would write equations on the napkins, just as we were doing. After that, the waiter took our order and went off.

Carlos Rivera Cruchaga[24] was born in 1925 in Santiago de Chile, where he studied, after spending three years of his youth in Germany, at the *Colegio*

---

[20]T. Regge's letter to E. Recami of November 28, 1978 in Recami (1987).

[21]C. Rivera's letter to E. Recami of October 18, 1978 in Recami (1987).

[22]C. Rivera's letter to M. Majorana of November 28, 1978 in Recami (1987).

[23]See the interview with Violeta Rivera in *Chi l'ha visto?*, a television programme broadcast in December 2006 on the Italian TV channel Raitre.

[24]The following information was taken from two short biographical reviews of Rivera prepared, respectively, by Jorge Ossadòn in 1999 and by Francisco Claro in 2004 (the year Rivera died). We thank Professor Claro for kindly making them available.

Sagrados Corazones de los Padres Franceses de Alameda. He graduated in 1943. After hesitating for a while, he continued his studies of mathematics and physics at the University of Chile, and in 1948 was already working as an assistant there. He subsequently obtained a scholarship to further his studies of physics in Europe, where he spent three years carrying out advanced research in optics in Spain, at the *Instituto de Optica Daza de Valdes*. He then specialised in neutron physics at the Max Planck Institute for Physics in Gottingen, Germany, directed by Heisenberg. However, these studies do not seem to have produced results that could be published in scientific journals (the first publication carrying Rivera's name dates from 1967 and deals with a secondary didactic issue of physics[25]). When he went back home he worked at the Pedagogical Institute of the University of Chile from 1955 to 1959 and, later on, until 1961, at the Universidad Austral, where he founded the first degree course in physics. From 1961 until his retirement in 1986 (when he was granted the title of *professor emeritus*), he went back to the Catholic University of Santiago, where he created the *Escuela de Fisica* inside the Department of Engineering. During his active life, Rivera devoted himself mainly to pedagogical studies for the scientific education of young people, and this brought him much honorary recognition from various institutions. In the 1980s, he also participated in the production of educational television shows.

Carlos Rivera's personal résumé, therefore, would not give reason to doubt his good faith regarding the claims he made about Majorana, as just reported. But on closer inspection, some of the inaccuracies in Rivera's recollections cannot be overlooked.

First of all, according to what Rivera himself reported in more detail in another interview,[26] his first stay in Buenos Aires before leaving for Europe must have happened "in January 1950", while in the recent interview with his wife Violeta, we are told that this trip took place in October of that year. According to what we know about his scholarship in Europe, we cannot reasonably set the episode with Mrs. Talbert in either of the quoted dates, but rather in March 1950.[27] It should be said, however, that this journey is known[28] not to be to Germany, but to Spain, where Rivera was to start working at the already mentioned *Instituto Daza de Valdes*. It was only at the end of 1951 that Rivera moved to Gottingen to work at the institute directed by Heisenberg, thanks to the director's interest in the Spanish institute, and here he fulfilled his dream of carrying out studies of theoretical physics.

Even the details concerning the acquaintanceship between Mrs. Talbert and the Rivera Cruchaga family are slightly contradictory. As a matter of fact, in contrast

---

[25]See the paper (Rivera, Infante and Claro 1967).

[26]Ponz de Leon, television programme, *loc. cit.*

[27]This detail is clearly of secondary importance, and different stories could perhaps live together if we make the plausible assumption that Rivera left several times for Europe from the Argentinian capital.

[28]Ossandòn, see *supra*.

with what was written in the letter to Maria Majorana, Rivera says in another interview:[29]

> My father had a friend in Buenos Aires, Mrs. Talbert. They had met in Europe, and when I went to study with Heisenberg in Germany, he asked me to go and visit Mrs. Talbert, to give her his regards.

On the other hand, the details of his second stay in Buenos Aires sound accurate. In January 1954 Rivera had to come back from Germany to South America, as the four-year study period covered by the scholarship had come to an end. We can also reasonably accept that even the third stay in the Argentinian capital is accurately reported, when Rivera went to the more than respectable Hotel Continental. According to Rivera's wife, that was in February 1961, a little more than a year after he had been appointed director of the Institute of Physics at the Universidad Austral of Chile.

Concerning the incident with Mrs. Talbert, we can no longer have direct confirmation, as there are no other direct witnesses. However, the following consideration regarding the consistency of Rivera's statements may be useful. It is known[30] that when Rivera attended the Pedagogical Institute in Santiago de Chile as a student, at the end of the 1940s, he took up the study of theoretical physics on his own using Gregor Wentzel's German textbook on quantum field theory (Wentzel 1943).[31] This book discusses the theory of nuclear forces in detail, and it appropriately highlights the model proposed by Majorana, which we have already talked about. So it might not be at all unreasonable to suppose that Rivera had Wentzel's text with him during the incident with Mrs. Talbert, and that he was studying Majorana's exchange forces when she came into the room.

One feature of the Rivera couple's final recollection, the one relating to what happened in the Hotel Continental, is both extravagant and hard to verify, namely, the claim that the waiter showed them Majorana's picture. But we should point out that Violeta Rivera's statement quoted above (from 2006) agrees well with what her husband Carlos had said back in the 1980s:[32]

> When I went to Argentina with my wife, we had lunch at the restaurant of the Continental Hotel. The maître d'hôtel, who already knew me as I had already been there on other occasions, said to me: "You are a physicist, you must know this person". He had a picture, and asked: "Do you know this gentleman?" "Ettore Majorana!" I replied. "Yes," he tells me

---

[29]Ponz de Leon, television programme, *loc. cit.*

[30]Ponz de Leon, television programme, *loc. cit.*

[31]One should not be surprised by the fact that Rivera was studying from a German book, given that he had been speaking the language since the long stay in Germany during his childhood. Wentzel's text might have been recommended by one of his professors of German origin who was teaching at the University of Chile: Erich Paul Heilmeier, director of the degree course in physics, who emigrated to Chile in 1937, or Kurt Reiseneger, a physical chemist who had worked in Berlin with Otto Hahn, Lise Meitner, and Fritz Strassmann, on the run from Nazi Germany because of his wife's Jewish origins.

[32]Ponz de Leon, television programme, *loc. cit.*

"there was a time when he would often come here to dine", before 1950. It was a picture of a much older Ettore, not like the one I had. [...] Here he was alone, older.

As we said before, there is actually no direct way to check this information, which could be of the utmost importance to establish beyond doubt Majorana's presence in Argentina. We know, from his partner Jolanda,[33] that the waiter's name was Baudano and that, before emigrating to Argentina in the 1940s, he was a mathematics teacher at a high school in Piedmont, Italy. On the other hand, although we have actually managed to trace a Baudano family in present day Piedmont, no confirmation has come from this about an emigrant in Argentina who belonged to that family.[34]

Clearly, the reported information is at least second-hand. Assuming the good faith of those involved, the usual rules of critical interpretation allow us to establish the following, with a certain degree of plausibility, leaving out those details that remain uncertain. Around the 1950s, an Italian physicist named Ettore Majorana was a friend of the engineer Tullio Magliotti, son of Mrs. Talbert, who also knew him, residing in Buenos Aires. Probably, in the same period, the same physicist frequented the Hotel Continental in that city, and was noticed by a waiter of the hotel. Most likely he did not live in Buenos Aires, but in some nearby town, such as Rosario or Santa Fè, and would come to the capital every now and then.[35] The other things Rivera reported, such as the mention of Fermi and the formulas written on the napkin, which would tend to link *that* Majorana with *our* Ettore Majorana, are not very convincing, because they may easily be later interpolations (even unintentional, and not necessarily Rivera's).[36] In this respect, it is useful to point out that, in 1950, there were already a lot of families in Buenos Aires with the name Maiorana or Maiorano (but not Majorana or Mayorana), as can be seen from the telephone directory of the time.[37] However, none of these families ever referred to an Ettore, and many if not all referred to people whose name was already Hispanic. After all, Rivera himself thought there might be a mix-up over names:[38]

At first it occurred to me that [Mrs. Talbert and I] were not talking about the same physicist. As there was a picture of Ettore Majorana together with a picture of Heisenberg in the book [which I was studying], I showed her the photo. "Oh yes," she said, "this is Ettore, but here he is very young." Of course, the picture dated back to 1936 or 1935.

---

[33]Ponz de Leon, television programme, *loc. cit.* See also S. Ponz de Leon's article in the Italian newspaper *La Repubblica,* September 30, 1987 and E. Recami's in *La Stampa,* April 1988.

[34]Information kindly made available by Mrs. Lucia Baudano.

[35]This is also the opinion of the headwaiter of the Hotel Continental at the time, according to what he told Ponz de Leon in the above-mentioned television programme (another second-hand testimony).

[36]An example of this is the fact that Rivera identifies Tullio Magliotti as "graduate in electrical engineering" at the beginning of his account, while later on he is already described as "the engineer Magliotti".

[37]This directory is kept, for example, at the Museo Postal y Telegrafos - Museo de Telecomunicaciones of the Correo Central of Buenos Aires.

[38]Ponz de Leon, television programme, *loc. cit.*

In Wentzel's book, there is no picture, either of Majorana or Heisenberg, but it seems quite probable that Rivera had other papers with him, such as the proceedings of conferences or workshops attended by the two scientists (most likely from 1933, rather than 1935–36, when Majorana was with Heisenberg in Leipzig). However, we have not been able to find any such volume containing a picture of Majorana.

There are other indications which would suggest Majorana's presence in Buenos Aires, and which are *independent* from Rivera's. But it does not seem possible to extract any reasonably reliable information from these, apart from the fact that they exist and, we reiterate, are independent of Rivera's information.[39]

On the Argentinian Trail

The lead arising from Rivera's testimony (and the others just reported) seems to point toward the possible presence of Majorana in Argentina. However, such indications *alone* are barely reasonable, and anyway difficult to validate explicitly. But if we also take due note of the other indications, which cannot be neglected and point independently in that direction, it would seem necessary and useful to follow up this lead. These new enquiries have now been carefully carried out.

First of all, it has not been possible to get much useful or detailed information about Mrs. Talbert who was probably from France. As a matter of fact, we do not even know her first name because, very likely, Frances Talbert was only her family name, following the practice in Spanish-speaking countries of using both the father's and the mother's surname. This would suggest that this lady had been in Argentina for some time prior to 1950. However, there is no user with this compound surname in the Buenos Aires telephone directories of 1950-51. There are four under the name Frances, but none is assigned to a woman.[40] The only case worth mentioning might be that of a "Frances family" living in Calle Nepper, 104b. In the outskirts of Buenos Aires there were then several telephone users with the name Frances, and three of them were women (Juana Frances, Maria M. A. Farm Frances, and Maria E. T. de Frances). However, all the Frances families at present in Buenos Aires and the surroundings bear no relation with our witness here.

No hotel, boarding house, or bed and breakfast run by a Frances appears in the telephone or business directories of the considered period (even the more detailed area directories). In 1951, there was a boarding house in Calle Bolivar 1645 whose owner was a certain Vicente Rozas. This establishment belonged, in 1965, to a Frances Anselmo. The Hotel de France in Calle S. Josè 9, in activity in Buenos

---

[39]This information is reported in Recami (1987); but see also *Chi l'ha visto?* television programme, *loc. cit.*

[40]Note that, according to Rivera's statement, Mrs. Talbert had a telephone line in her boarding house.

Aires since 1950, does not look like Rivera's description of a "boarding house". Even if it could be linked to some owners of French origin, there would not appear to be a connection with a Mrs. Frances Talbert. The only hotel at present in service in the whole of Argentina which might be of some interest is the Frances Hotel in Mar del Plata, a city about 400 km from Buenos Aires and a haven for many Italian emigrants. This hotel was sold to its present owner by Lionor Hugalde de Frances in 1986. But even here, no plausible link can be established between this hotel and our Mrs. Talbert.

Therefore, it seems reasonable to say that the boarding house run by Mrs. Talbert was not officially registered with the competent authorities, and this may have been the reason for Mrs. Talbert and Magliotti's sudden "disappearance", without necessarily appealing to any more imaginative explanation.

On the other hand, as far as Tullio Magliotti is concerned, he was certainly of Italian origin like Majorana, and we have been able to establish the following.

To date, there are five Magliotti families in Italy (four of them listed in the telephone directory), all descending from a single lineage in Liguria. However, none of these has been able to provide any information about the engineer Tullio Magliotti, whom they clearly did not know. But this may be because their forefather had long been dead when present members of the Magliotti family were still very young, whence their sources would inevitably be second-hand. Nevertheless, we did discover that their forefather's brother, Antonio, had emigrated abroad for financial reasons much earlier, either to Africa or Latin America. Given the age of today's Magliottis, it is just possible that Antonio Magliotti was the father of the engineer Tullio. In any case, it is interesting to note that there are at least 17 Magliotti families across the whole of Argentina today, far more than in Italy; and in the Buenos Aires telephone directory, there were already three users in 1950 (although this last piece of information is of limited interest, in contrast to Mrs. Frances Talbert's case, since we have already noted that she must have owned a telephone contract in her name).

As regards Tullio Magliotti's profession, "ingeniero mecanico y eletricista" (mechanical and electrical engineering), we can only say that in the Buenos Aires telephone and business directories of 1950–51 there is no "Magliotti" under the heading "Ingenieros". No other information has so far been found since no other registers or documents are currently available.

To be thorough, we should mention the presence in the Argentinian capital, since the years we are interested in here, of families with the names Magliotte, Magliotto, Gagliotti, and similar, which are even more numerous than the Magliottis, although it has not been possible to find any Tullio.

Clearly, this body of information would tend to confirm the plausibility of Rivera's statements, and there would thus appear to be some truth in his story (provided we assume, as we do here, that the witness spoke in good faith).

The last piece of information in Rivera's statement, the one least likely to be realistic, hence the one that needs to be most carefully checked, is the story about the Hotel Continental, which at first glance appears to be unrelated to Frances Talbert and Tullio Magliotti.

But it is just this piece of news which comes as a surprise...

We have already seen that Rivera's stay at the Hotel Continental in 1960 was highly plausible. Ignoring details that would be difficult to check, Rivera's information is that a physicist named Ettore Majorana frequented that hotel in the 1950s. It is a large hotel in Buenos Aires, comprising a single building located at the corner of two streets, detached from the buildings nearby, with the address 725 Avenida R. S. Pena. In the period that interests us here, it had a large restaurant on the ground floor, perhaps with an independent entrance, at the corner of the building. This would have been the scene of the events described by Rivera. The Buenos Aires business directory from 1951 shows as many as 324 hotels in the city, and the addresses of 77 of these are highlighted to distinguish them from the others in the directory; but strangely, the very large Hotel Continental is not among these 77.

Even if we acknowledge, as a working hypothesis that could not be checked directly, that our Majorana actually frequented the Hotel Continental restaurant, the obvious question would be: *why* did Majorana go to that restaurant and *why* was he noticed by a waiter who, according to Rivera, spotted Majorana writing down scientific equations?

As far as these questions are concerned, the enquiries carried out on site *only* ascertained the surprising proximity of the Hotel Continental to 222 Calle Perù, which, in 1950–51, was the building housing the Faculty of Exact, Physical, and Natural Sciences of the University of Buenos Aires (together with the *Academia Nacional de Ciencias Exactas, Fisicas y Naturales*).

The first reason why this coincidence is so striking is that, in those days, it was precisely this faculty that conferred the engineering degree. If the elusive Tullio Magliotti had graduated in Buenos Aires, he would have had to obtain his degree in the above-mentioned *Facultad de Ciencias Exactas, Fisicas y Naturales*. In any case, it would certainly have been strange if the engineer Magliotti had not known the place. The reasonableness of this first observation would suggest an underlying connection between the different pieces of information reported by Rivera (regarding Magliotti and the Hotel Continental), a connection that Rivera could not have intended, since he was probably unaware of this.

The second reason is even more surprising. As we said in Chap. 8, the little folder with Majorana's notes for his Naples lectures, which he gave to one of his students before disappearing, probably contained a manuscript with some personal research notes written in French. Regarding the date when it was written, we speculated that it might have related to a series of conferences held by Enrico Fermi in Buenos Aires in 1934. Surprisingly, as we said, the place where Fermi gave his conferences that August was precisely the *Facultad de Ciencias Exactas, Fisicas y Naturales* of the University of Buenos Aires... (Fermi 1934). According to Segrè's recollections, once he was back in Rome, Fermi spoke enthusiastically about his visit to South America to his friends and colleagues in the Rome group: "Fermi gave his lectures in Italian before a very large audience and was very pleased with the interest in his research" (Segrè 1970). So it is quite conceivable that Majorana had known about these details at least since 1934. Note also that, as far as we know, the only copy in Italy of Fermi's Buenos Aires lectures is not kept in Rome, but at the university library in Bologna. It came there from the Royal Academy of Sciences of the Institute, of which Ettore's uncle, the experimental physicist Quirino Majorana, was a member. Actually, we may suppose that Majorana's knowledge of the Buenos Aires *Faculdad* dated back some time earlier. In fact, in 1928, through the mediation of the Argentinian Institute of Italian Culture,[41] the famous mathematician and epistemologist Federigo Enriques, father of Giovanni, a classmate of Majorana's, was invited to give some lectures there. Amaldi (1968) remembers that "in 1928, in May and June, exam time at the university, we used to meet before dinner, between seven and eight in the evening, at the *Casina delle Rose* of the *Villa Borghese*. Ettore Majorana, Giovanni Gentile Jr, Emilio Segrè, and I from the Institute of Physics were joined by Luciano Majorana, Giovanni Enriques, Giovanni Ferro-Luzzi, and Gastone Piqué, all students of engineering in the same year as Ettore." So, in this or in other circumstances, the young Giovanni Enriques must have talked about his father's visit to Buenos Aires. But these were not the only two opportunities for Ettore to hear about the *Facultad de Ciencias Exactas* of the Argentinian capital. As a matter of fact, he might have heard about it in 1930, when Francesco Severi, one of Majorana's professors, also gave some lectures there. In one of these, among other things, he even spoke about the work of the young physicist Enrico Fermi (Severi 1931). But above all, in 1937, some months prior to Ettore's disappearance, another of his esteemed professors, Tullio Levi-Civita, was invited for the same reason to the same institute.

The relationship between the mathematicians in Rome and those in Buenos Aires, especially Enrique Butty and Julio Rey Pastor, is also confirmed in some texts published by the Argentinian *Facultad* and sent to the Royal School of Engineering library in Rome (today they are kept in the library of the Department of Mathematics "Guido Castelnuovo" of Sapienza University in Rome). Some of these texts, as is easily checked, were given to this library by Federigo Enriques and Francesco Severi. Another, particularly intriguing one is the catalogue of the library of the Buenos Aires *Facultad*.[42] From this catalogue we learn that it was a well-stocked library, both for science books and for specialised magazines, as they held practically every important physics journal there, including *Annales de Physique, Nature, Il Nuovo Cimento, Proceedings of the Royal Society of London, Rendiconti dell'Accademia Nazionale dei Lincei*, and *Zeitschrift für Physik*. But we would also like to mention the curious presence in the catalogue of certain secondary journals, such as the reports of the *Accademia delle Scienze Fisiche e Matematiche* and of the *Annali del R. Istituto d'Incoraggiamento*, as well as the *Atti del Collegio degli Ingegneri e Architetti*, all published in Naples.

---

[41]The points discussed here can be found in the following texts: *Ciclo de conferencias científicas y de carácter general. Tomo III*, Sociedad Científica Argentina, Buenos Aires, 1945; L. A. Santaló et al., *Evolucìon de las ciencias en la Repùblica Argentina 1923–1972. Tomo I. Matemàtica*, Sociedad Científica Argentina, Buenos Aires, 1972.

[42]Facultad de Ciencias Exactas, Fisicas y Naturales, *Catalogo de la Biblioteca*, University of Buenos Aires, Buenos Aires, 1931.

The peculiarity of the catalogue in Rome referred to above is that it is incomplete. In fact, the first 256 pages are missing. This might have happened for some trivial reason, but we cannot help noticing that in the missing part there were references to almost every mathematical, physical, and even philosophical text that Majorana might have considered interesting: for example, the books by Hilbert, Poincaré, Rutherford, Sommerfeld, Weyl, and Whittaker. We are not, of course, suggesting that Majorana was involved removing these pages. But it does not seem totally unreasonable to suggest that someone who frequented libraries as assiduously as Ettore would have ascertained the relevance of the *Facultad de Ciencias Exactas, Fisicas y Naturales* in Buenos Aires as a place of study and research, just by referring to this catalogue.

The last amazing fact regarding the *Facultad* is a note on the books owned by the library at the institute around 1950, possibly the time of Rivera's story about Magliotti. As already mentioned, Majorana's personal library was rather limited, consisting of only 29 texts. But 19 of these texts were (and still are) available at the *Facultad*, 9 of them in the same edition as the one owned by Majorana, while for the others there were different editions (some of these texts, even published before 1931, are not in the above-mentioned 1931 catalogue, and were bought later on). Apart from those which might reasonably be found in any "ordinary" scientific library, there were fundamental texts on group theory,[43] surely appreciated by Majorana, but at the same time given little consideration by most physicists in those early days, and there were the books in Italian by Fermi (1928), Rasetti (1936), and Persico (1936).

A Careful Analysis

Everything known to us and considered relevant regarding Majorana's time in Naples and his subsequent disappearance has been carefully presented above; we have only left out things which cannot be considered certain, or have not been confirmed later. Always with due caution, we have discussed the data and even the more *imaginative* elaborations on them, at least those which cannot be completely rejected out of hand, and have tried to extract the most plausible and useful information. So at the end of this presentation, what can we conclude about Majorana's disappearance? The main data available, on which we shall focus our attention, is given in the following list, where an asterisk indicates uncertain or not fully confirmed data. Note that some items may not be directly linked to the disappearance but, treated with reasonable caution, should probably not be neglected.

---

[43]That is, (Bianchi 1928), (Speiser 1927), (Weyl 1928).

1. January 13: Majorana begins the theoretical physics course in Naples.
2. January 18: Occhialini arrives in Naples from Sao Paulo on the ship *Oceania*, and meets Majorana at the Institute of Physics.
3. January 27, 29: in his lecture notes, Majorana makes four references to future lessons, never actually given.
4. February–March: Majorana withdraws three and a half months of his teaching salary; if we subtract his expenses in Naples and add the total taken from his personal bank account (sent to him from Rome by his brother), the result, amounting to not less than 10,000 Italian liras, is not found, and neither is his passport, after the disappearance.
5. March 22, 24: in his lecture notes, Majorana makes three references to future lessons, never actually given.
6. March 25, morning: Majorana gives the lecture notes to G. Senatore, who was born in Sao Paulo, Brazil.
7. *The folder with the notes also contains some sheets of paper with personal research notes, which can perhaps be indirectly linked to E. Fermi's visit to Buenos Aires in 1934.
8. March 25, at about 05:00 p.m.: Majorana leaves the *Albergo Bologna*.
9. March 25: Majorana writes the letter "to my family" (left in his hotel room) and sends a farewell letter to Carrelli; both letters express suicidal thoughts.
10. In the letter to Carrelli, Majorana sends his regards to S. Sciuti, another student of his, but not to Senatore.
11. *March 25, 10:30 p.m.: Majorana allegedly embarks on the ferry *Città di Palermo* to cross to Palermo (but the ticket receipt has never been made available to the investigators).
12. March 26: a telegram signed by Majorana is received by Carrelli from Palermo, and another is sent to the management at the *Albergo Bologna*; in these, and in a letter to Carrelli written on the letterhead of the *Grand Hotel Sole* in Palermo, he seems to be having second thoughts about suicide.
13. *(Not dated): Majorana may have warned his cousins in Catania that he was in Sicily, but that he could not go and visit them; in Palermo he may have looked for Segrè at the Institute of Physics, without finding him.
14. *March 26, 11:00 p.m.: Majorana allegedly embarked on the ferry *Città di Palermo* to cross to Naples (but the ticket receipt has never been shown to the investigators).
15. *V. Strazzeri, on board that ferry, allegedly recognises Majorana as being one of his two cabin mates.
16. March 26: from the port of Naples, the ship *Oceania* leaves for Buenos Aires (the same ship that brought Occhialini to Naples). In the Naval Almanac of that year, which Majorana owned, there is a picture of this ship, the only cruise ship to be illustrated in the almanac.
17. Majorana disappears at the weekend; the enquiries begin four days later.
18. *(Not dated): Carrelli suggests that Majorana was involved in spiritual exercises at a convent in Naples (a highly unreliable testimony: glaring inaccuracy).

19. (Not dated): a religious lead is explored (with alleged sightings), but nothing absolutely certain can be concluded.
20. *First days of April: a nurse claims to have seen Majorana in Naples (a highly unreliable testimony: glaring inaccuracy).
21. The investigators consider that the Majorana case must effectively be "filed" as suicide.

From this, it is clear that, whatever else can be said, Majorana's fate was neither accidental nor unintentional, but prepared in advance and in some detail.

The conclusion drawn by the investigators that Majorana committed suicide (shared by his friends and most of his relatives, with the exception of his mother), is by far the easiest and most reasonable, but it seems to contradict several pieces of information. And it would perhaps be unreasonable to consider *all of them* as accidental. But even if we assume that they were, we should at the same time suppose that Majorana's detailed preparation (clear from the above) was meant to distract attention as far as possible from this tragic conclusion. Frankly, though, this would seem excessive if the explanation is that he wished to alleviate his mother's pain and give her hope.

The religious lead, which also stands on frail grounds, looks like a clever red herring if, as it seems reasonable to think, it was instigated unconsciously by Carrelli, who was of course informed by his friend Majorana.

If he did "prepare" things in this way, he would probably not have been able to take into account the superficiality with which the Naples police (and not only them) carried out their enquiries. But how else can we interpret so many testimonies, each containing its own elements of doubt? And it is also amazing that the only piece of evidence for Majorana's voyage to Palermo is a sheet of headed notepaper from a hotel there, and the good faith of those carrying out the investigation.

Many clues, some of which were given indirectly by Majorana himself, seem to suggested, in a rather pointed way, to a lead abroad, in South America. At the present time, the main data that would appear to confirm this are as follows, with the usual proviso that, even if confirmed, they may not actually be relevant to the Majorana case.

22. March 1950: while a guest at Mrs. Frances Talbert's boarding house in Buenos Aires, Carlos Rivera is studying Majorana's model of nuclear forces, and his landlady tells him that her son, Tullio Magliotti, is a friend of the physicist Ettore Majorana.
23. *The engineer Tullio Magliotti allegedly met Majorana in his capacity as an engineer, though he knew that he was originally a physicist.
24. The institute in Buenos Aires that trains engineers is the *Facultad de Ciencias Exactas, Fisicas y Naturales* of the University, which in the 1930s had hosted some of Majorana's colleagues and acquaintances: F. Enriques (1928), F. Severi (1930), E. Fermi (1934), T. Levi-Civita (1937).

25. *Majorana knew, before his disappearance, about the activities of this *Facultad*, and about the existence of a very well stocked library there, which would be useful for carrying out research in the field of theoretical physics.

26. *February 1961: a waiter at the Hotel Continental in Buenos Aires tells Rivera that the physicist Ettore Majorana used to frequent the hotel restaurant in the 1950s.

27. The Hotel Continental is near the *Facultad de Ciencias Exactas, Fisicas y Naturales*.

28. *(Not dated): other testimonies,[44] independent from Rivera's, also suggest Majorana's presence in Buenos Aires.

Actually, this seems like an accumulation of curious coincidences but, as we said above, it does seem surprising to think that they could all have occurred by chance. But can these clues, as such, be taken as the proof of an ingenious and successful plan? Of course, it is surprising that such a lead was not followed up by the investigators in 1938, despite the fact that they could easily have checked at the port of Naples (ship arrivals and departures); this only emerged later on and in a way that was totally independent from the clues that should have generated it.

And now our analysis must end. It is up to the reader to develop his/her own idea of what has been presented.

"As Fermi observed, with his intelligence, if he had decided to disappear or to have his body disappear, Majorana would certainly have had no problem doing that" (Amaldi 1968).

And, as a matter of fact, he did so.

Majorana's *plot*, like his deepest physical theories, has remained impenetrable until today, perhaps hiding from view the deep solution to an enigma.

---

[44]Reported in Recami (1987).

Epilogue

So what actually became of Majorana?

The answer is simple, even trivial: Majorana is alive. In his daring ideas and intuitions, which are still under both the theoretical and experimental spotlight. In his calculations, the basis of so many applications. In his theories, stored away in notebooks kept for experts, still waiting to come to light and be made known. In light of this, the spirit of the physicist, lively and critical, cannot be said to have gone. The disappearance of the person Majorana is, on the other hand, something quite different, although secondary compared to his talent, which remains right here with us. Perhaps we could say that our commitment in the search, to establish his fate, is justified by the prospect of his legacy to us, apart from what he has already abundantly entrusted to us.

In the previous pages, we have discussed both his talent and the mystery surrounding him. The framework of clues we have provided regarding Majorana's disappearance and possible subsequent events speaks for itself and no further comment is needed. Certainly, some of these circumstances would be better explained as the result of pure chance than as a skilful construction, even if a brilliant one. But the curious reader who has stayed with us up to now cannot be denied *another* story, one that can only be written once we shed the scruples of the historian and the scientist.

This story begins on January 10, 1938. In his Rome residence where he lives with his mother and siblings, EM receives the official letter appointing him to the tenure at the University of Naples. He quickly prepares everything he needs to move to Naples, where he can at last make his deep knowledge of theoretical physics available to young students, and where he thinks he may be able to loosen the ties that bind him to his mother, with whom he has so often argued, but whom he loves so deeply. Unfortunately, only his longing for teaching is crowned with success, despite the lack of preparation and ability of his students. As a matter of fact, EM's mother not only refuses his request not to come to Naples with other relatives to attend the opening lecture of the course, but continues to exercise her unbridled motherly influence. So the hotel he chooses is not good for him: he has to move, and move again, to find one which is more appropriate. And then he needs a nurse to assist him, because of his delicate health. And so on and so forth, until she

© Springer International Publishing AG 2017
S. Esposito, *Ettore Majorana*, Springer Biographies,
DOI 10.1007/978-3-319-54319-2

even goes as far as to envisage a move to Naples: a new house must be found, a new national health registration, documents for the new residence, ...

Meanwhile, on January 18, EM receives a brief visit from a physicist colleague of his, one he has never met before, coming from Sao Paulo in Brazil, who tells him about his life there. EM immediately thinks about his beloved professor of rational mechanics, who just months earlier had also crossed the ocean to give lectures in Buenos Aires. But a few years earlier, his colleague and friend, and former supervisor for his physics degree, had also gone to the same Argentinian university, and had spoken well of: the people were welcoming, there was a well-stocked library, and the situation was much more peaceful than in Italy... EM even says a little too much to the person visiting him on that January 18, and this person does not at first pay due attention (how could he?), but then *misunderstands*. With his sharp mind, EM notices that, of course. And what if that was the ultimate solution for *his* life? It would be necessary to implement a careful plan of action, but the skills and means were surely not wanting. A nice piece of theoretical work, carefully carried out, that would get people believing *everything* and *nothing* at the same time.

First of all, however, nothing must leak out. So the following week-end, Saturday January 22, EM goes back to Rome to get his student books, in order to prepare some notes to give to his pupils. The course is going to be a long one, so it's better to explain, with appropriate references to future lessons, that they will need to work hard. On the other hand, such a plan will require plenty of money. This is no big problem, but he needs to act quickly, and asks his brother in Rome to withdraw everything from his bank account and send to him: in one week, on January 29, EM will be able to collect his salary, and the request for more money may arouse some suspicions.

The first ship to South America, the same one that had brought his colleague to visit him in Naples, would set sail on February 6: EM could plan everything for that date, but it is better not to hurry. It is better to prepare everything properly, and moreover, if he could withdraw another salary, that would be helpful. No, the right date is March 26. That way there is plenty of time to draw up a very detailed plan, so that those who carry out the enquiries will conclude that it was a suicide (even if his body is never found), and also to scatter around plenty of clues pointing in quite different directions, both to divert suspicions and to give hope to any who may mourn his disappearance.

Meanwhile, to make his plan work perfectly, he must prepare the ground for the one who will unwittingly play the key role: the director of the institute he teaches in, who has by now become a close friend of his. He will be the one who spreads the news of his disappearance, but not immediately; and he will also be the one who will suggest the direction the enquiries should take. So EM pretends he is feeling strange and exhausted; he wishes to perform spiritual exercises in some convent, to recover his lost peace of mind. And then, as fate would have it, there is a unique opportunity: a week before the day chosen for his disappearance is Saint Joseph's day. There are no lectures at the university, and he has four days (from Friday, March 18, to the following Monday) to set his plan in motion; he will say goodbye to his family some weeks earlier. EM goes to Palermo and prepares the scenario that will take place in a week. He hires a boy to have a letter and a telegram sent to his director, to arrange for his hotel room to be kept in Naples, and prepares everything in an appropriate hotel. Of course, he cannot expect fate to assist him again with the matter of the return ferries he *will not be able* to catch. But something can still be worked out with his cousins living in Sicily and a mathematician who is a friend of one of his former student friends, who now teaches in Palermo.

The plan is ready: it is March 25. EM even has time to leave his legacy, his lecture notes, to one of his students; in fact, the very student who was born in South America. Why not leave a few helpful indications among his gestures and written papers? They will be dispersed among so many other carefully placed clues: that is the best way not to reveal the hidden truth. The rest of his legacy will be found in the hotel where he is staying.

The plan is finally carried out. Any last-minute hesitation? Any second thoughts? Would it be better to put everything off? Or even just postpone (seeking refuge, in the meantime, in the Mental Hospital)? It does not matter. Eventually, everything works out all right, and EM is where he wants to be, in Buenos Aires.

This is what the more imaginative reader may have pictured in his/her mind; perhaps just a nice piece of fiction. There is no irrefutable evidence that this EM is *our* Ettore Majorana.

A decisive piece of evidence. As it would be the presence of our protagonist in the 1938 disembarkation list in Argentina, still kept at the Buenos Aires *Direccion Nacional de Migraciones*. A piece of evidence that has been sought, but never found. Majorana's name does not appear among the 587 passengers of the "Oceania", who disembarked at the *Darsena Norte* of the Argentinian capital on April 11, 1938. A different name, another ship...?

Is the Majorana case destined to become a *cold case*? Perhaps.

References

Ackermann, J., & Hogreve, H. (1991). Adiabatic calculations and properties of the He_2^+ molecular ion. *Chemical Physics, 157*, 75.

Amaldi, E. (1966). *La vita e l'opera di E. Majorana*, Accademia Nazionale dei Lincei, Rome. English edition: Amaldi, E. (1966). *Ettore Majorana: Man and scientist*, In Zichichi, A. (Ed.). *Strong and Weak Interactions*, New York: Academic Press.

Amaldi, E. (1968). Ricordo di Ettore Majorana. *Giornale di Fisica, 9*, 300.

Anderson, C. (1932). The apparent existence of easily deflectable positives. *Science, 76*, 238.

Arimondo, E., Clark, C. W., & Martin, W. C. (2010). Ettore Majorana and the birth of autoionization. *Reviews of Modern Physics, 82*, 1947.

Bartocci, C., et al. (2007). *Vite matematiche—Protagonisti del '900 da Hilbert a Wiles*. Milan: Springer.

Barut, A. O., & Duru, I. H. (1973). Introduction of internal coordinates into the infinite-component Majorana equation. *Proceedings of the Royal Society of London, A 333*, 217.

Barut, A. O., & Nagel, J. (1977). Interpretation of space-like solutions of infinite-component wave equations and Grodsky-Streater 'No-Go' theorem. *Journal of Physics A: Mathematical and General, 10*, 1233.

Bassoli, R. (1998). "Prodigio di famiglia", in *L'Unità*, January 27.

Bethe, H. A. (1955). Memorial symposium held in honour of Enrico Fermi at the Washington meeting of the American Physical Society, April 29, 1955. *Reviews of Modern Physics, 27*, 249.

Bhabha, H. J. (1945). Relativistic wave equations for the elementary particles. *Reviews of Modern Physics, 17*, 200.

Bianchi, L. (1928). *Lezioni sulla teoria dei gruppi continui finiti di trasformazioni*. Bologna: Zanichelli.

Bitter, F. (1962). Magnetic resonance in radiating or absorbing atoms. *Applied Optics, 1*, 1.

Blackett, P. M. S., & Occhialini, G. P. S. (1933). Some photographs of the tracks of penetrating radiation. *Proceedings of the Royal Society of London, A 139*, 699.

Black, F., & Scholes, M. (1973). The pricing of options and corporate liabilities. *The Journal of Political Economics, 81*, 637.

Blatt, J. M., & Weisskopf, V. F. (1952). *Theoretical Nuclear Physics*. New York: Wiley.

Bloch F., & Rabi, I. I. (1945). Atoms in variable magnetic fields. *Reviews of Modern Physics, 17*, 237.

Boniolo, G. (1987). Non prevedibilità e modelli sociologici desunti dalle scienze formalizzate. In Antiseri D., Infantino L., & Boniolo G. (Eds.). *Autonomia e Metodo del Giudizio Sociologico*. Rome: Armando Editore.

Bonolis, L. (2008). *Maestri e allievi nella fisica italiana del Novecento*. Pavia: La Goliardica Pavese.

Bontems, V. (2013). Ettore Majorana's transversal epistemology. *Revue de Synthèse, 134*, 29.

Boyanovsky, D. (1989). Field theory of the two-dimensional Ising model: Conformal invariance, order and disorder, and bosonization. *The Physical Review, B 39*, 6744.

© Springer International Publishing AG 2017
S. Esposito, *Ettore Majorana*, Springer Biographies,
DOI 10.1007/978-3-319-54319-2

Brossel, J., & Bitter, F. (1952). A new "Double Resonance" method for investigating atomic energy levels. *The Physical Review, 86,* 308.

Calvesi, M., & Corsi, A. (1989). *Giuseppe Sciuti.* Nuoro: Ilisso.

Carrelli, A. (1925). Sul fenomeno di Tyndall. *Rendiconti della R Accademia Nazionale dei Lincei, 1,* 279.

Carrelli, A. (1928). Sulla relatività a cinque dimensioni. *Rendiconti della R Accademia Nazionale dei Lincei, 7,* 566.

Carrelli, A. (1940). *Lezioni di Fisica Teorica. La teoria di relatività.* Naples: G.U.F.

Carrelli, A. (1946). Sulla polarizzazione della luce del cielo. *Rendiconti della R. Accademia Nazionale dei Lincei, 1,* 493, 907, 1012, 1242.

Castellani, L. (1974). *Dossier Majorana.* Milan: Fabbri.

Castellano, C., Fortunato, S., & Loreto, V. (2009). Statistical physics of social dynamics. *Reviews of Modern Physics, 81,* 591.

Castelnuovo, G. (1956). *Lezioni di geometria analitica.* Rome: Società Editrice Dante Alighieri.

Clementi, E., & Corongiu, G. (2007). Merging two traditional methods: The Hartree–Fock and the Heitler–London and adding density functional correlation corrections. *Theoretical Chemistry Account, 118,* 453.

Cohen-Tannoudji, C. (2003). Lecture course on "Interactions atomes-photons: bilan et perspectives" held at the College de France.

Cornell, E. A., & Wieman, C. E. (2002). Bose-Einstein condensation in a dilute gas: The first 70 years and some recent experiments. *Reviews of Modern Physics, 74,* 875.

Corongiu, G. (2007). The Hartree-Fock-Heitler-London method, III: Correlated diatomic hydrides. *The Journal of Physical Chemistry, A 111,* 5333.

Coman, L., et al. (1999). First measurement of the rotational constants for the homonuclear molecular ion He_2^+. *Physical Review Letters, 83,* 2715.

Condon, E. U., & Shortley, G. H. (1935). *Theory of atomic spectra.* Cambridge: Cambridge University Press.

Coulson, C. A., & Fischer, I. (1949). Notes on the molecular orbital treatment of the hydrogen molecule. *Philosophical Magazine, 40,* 386.

Della, Seta F. (1996). *L'incendio del Tevere.* Udine: Gasparri editore.

De Gregorio, A. (2007). Il 'protone neutro', ovvero della laboriosa esclusione degli elettroni dal nucleo. *Physis, 44,* 153.

De Gregorio, A., & Esposito, S. (2006). A lezione dal Professor Majorana. *Sapere, 72*(3), 56.

De Gregorio, A., & Esposito, S. (2007). Teaching theoretical physics: The cases of Enrico Fermi and Ettore Majorana. *American Journal of Physics, 75,* 781.

De Mauro, M. (1965). I ragazzi di Corbino. *L'Ora* (Palermo), October 6.

Di Grezia, E., & Esposito, S. (2004). Fermi, Majorana and the statistical model of atoms. *Foundations of Physics, 34,* 1431.

Di Grezia, E. (2006). Majorana and the investigation of infrared spectra of ammonia. *Electronic Journal of Theoretical Physics, 3,* 225.

Dirac P. A. M. (1928a). The quantum theory of the electron. *Proceedings of the Royal Society of London, A 117,* 610.

Dirac P. A. M. (1928b). The quantum theory of the electron. Part II. *Proceedings of the Royal Society of London, A 118,* 351.

Dirac, P. A. M. (1930). *The principles of quantum mechanics.* Oxford: Clarendon Press.

Dirac P. A. M. (1931). Quantised singularities in the electromagnetic field. *Proceedings of the Royal Society of London, A 133,* 60.

Dirac, P. A. M. (1972). A positive-energy relativistic wave equation. II. *Proceedings of the Royal Society of London, A 328,* 1.

Drago, A., & Esposito, S. (2007). Ettore Majorana's course on theoretical physics: A recent discovery. *Physics in Perspective, 9,* 329.

Dragoni, G. (2006). Ettore Majorana as a guide in Quirino Majorana's experiments. Original letters and documents on an experimental and theoretical collaboration. *Proceedings of the Conference "Ettore Majorana's Legacy and the Physics of the XXI century"*, PoS(EMC2006) 005.

Dragoni, G. (Ed.) (2008). Ettore e Quirino Majorana—Tra fisica teorica e sperimentale. Rome-Bologna: CNR-SIF.

Esposito, S. (2002a). Majorana solution of the Thomas-Fermi equation. *American Journal of Physics, 70*, 852.

Esposito, S. (2002b). Majorana transformation for differential equations. *International Journal of Theoretical Physics, 41*, 2417.

Esposito, S. (2005a). Again on Majorana and the Thomas-Fermi model: A comment to physics/0511222, e-print arXiv:physics/0512259.

Esposito, S. (2005b). Il corso di Fisica Teorica di Ettore Majorana: il ritrovamento del documento Moreno. *Il Nuovo Saggiatore, 21*, 21.

Esposito, S. (2006a). A peculiar lecture by Ettore Majorana. *European Journal of Physics, 27*, 1147.

Esposito, S. (2006b). *Ettore Majorana: Lezioni di Fisica Teorica*. Naples: Bibliopolis.

Esposito, S. (2007a). Hole theory and Quantum Electrodynamics in an unknown manuscript in French by Ettore Majorana. *Foundations of Physics, 37*, 956.

Esposito, S. (2007b). An unknown story: Majorana and the Pauli-Weisskopf scalar electrodynamics. *Annalen der Physik, 16*, 824.

Esposito, S. (2012). Searching for an equation: Dirac, Majorana and the others. *Annals of Physics, 327*, 1617.

Esposito, S. (2014). *The physics of Ettore Majorana: Phenomenological, theoretical and mathematical*. Cambridge: Cambridge University Press.

Esposito, S., & Naddeo, A. (2012). Majorana solutions to the two-electron problem. *Foundations of Physics, 42*, 1586.

Esposito, S., Majorana, E., Jr., van der Merwe, A., & Recami, E. (Eds.). (2003). *Ettore Majorana: Notes on theoretical physics*. Dordrecht: Kluwer-Springer.

Esposito, S., Recami, E., van der Merwe, A., & Battiston, R. (Eds.). (2008). *Ettore Majorana: Unpublished research notes on theoretical physics*. Heidelberg: Springer.

Fano, U. (1935). Sullo spettro di assorbimento dei gas nobili presso il limite dello spettro d'arco. *Il Nuovo Cimento, 12*, 154.

Fano, U. (1961). Effects of configuration interaction on intensities and phase shifts. *The Physical Review, 124*, 1866.

Fermi, E. (1928). *Introduzione alla Fisica Atomica*. Bologna: Zanichelli.

Fermi, E. (1934a). *Conferencias*. Buenos Aires: University of Buenos Aires.

Fermi, E. (1934b). Tentativo di una teoria dei raggi β. *Il Nuovo Cimento, 11*, 1.

Fermi, E. (1937). Un maestro: O. M. Corbino. *Nuova Antologia, 72*, 313.

Fermi, E. (1960). *Collected papers*. Rome-Chicago: Accademia Nazionale dei Lincei—The University of Chicago Press.

Fermi, E., & Segrè E. (1933). Sulla teoria delle strutture iperfini, *Reale Accademia d'Italia, Memorie della classe di scienze fisiche, 4*, 131; reprinted in (Fermi 1960).

Ferrieri, G., & Magnano, A. (1972). "Il mistero Majorana". *L'Europeo*, May 11.

Fierz, M., & Pauli, W. (1939). On relativistic wave equations for particles of arbitrary spin in an electromagnetic field. *Proceedings of the Royal Society of London, A 175*, 211.

Fiori, G. (1971). L'atomica a Mussolini? Meglio sparire. *Tempo Illustrato*, November 28.

Fradkin, D. (1966). Comments on a paper by Majorana concerning elementary particles. *American Journal of Physics, 3*, 314.

Fukuda, Y., et al. (1998). Evidence for oscillation of atmospheric neutrinos. *Physical Review Letters, 81*, 1562.

Furry, W. (1938). Note on the theory of the neutral particle. *The Physical Review, 54*, 56.

Gentile, G. (1940). Sulle equazioni d'onda relativistiche di Dirac per particelle con momento intrinseco qualsiasi. *Il Nuovo Cimento, 17,* 5.

Gentile, G. (1942). *Ricordi di Giovannino.* Milan: unpublished.

Giannetto, E. (1993). Ettore Majorana e gli stati ad energia negativa. *Giornale di Fisica, 34,* 151.

Guarino, C. (1950). Ancora misteriosa la fine dello scienziato Ettore Majorana. *La Nuova Stampa,* July 29.

Heisenberg, W. (1931). Zum Paulischen Ausschließungsprinzip. *Annalen der Physik, 10,* 888.

Heisenberg, W. (1932a). Über den Bau der Atomkerne. I, in *Zeitschrift für Physik, 77,* 1.

Heisenberg, W. (1932b). Über den Bau der Atomkerne. II. *Zeitschrift für Physik, 78,* 156.

Heisenberg, W. (1933). Über den Bau der Atomkerne. III. *Zeitschrift für Physik, 80,* 587.

Heitler, W., & London, F. (1927). Wechselwirkung neutraler Atome und homöopolare Bindung nach der Quantenmechanik. *Zeitschrift für Physik, 44,* 455.

Kruger, P. G. (1930). New lines in the arc and spark spectrum of helium. *The Physical Review, 36,* 855.

Landau, L. (1932). Zur Theorie der Energieubertragung. *Physikalische Zeitschrift der Sowjetunion, 1,* 88 and *2,* 46.

Latora, V. (2005). Reti small world: l'architettura di un sistema complesso. *Il Nuovo Saggiatore, 21,* 77.

Lee, P. S., & Wu, T.-Y. (1997). Statistical potential of atomic ions. *Chinese Journal of Physics, 35,* 742.

Levi-Civita, T., & Amaldi, U. (1923). *Lezioni di Meccanica Razionale.* Bologna: Zanichelli.

Libonati, A. (1966). La verità sul primo 'giallo' atomico. *Gente,* July 6.

Majorana, E. (1931). I presunti termini anomali dell'Elio. *Il Nuovo Cimento, 8,* 78.

Majorana, F. (2007). Come io vedo Ettore Majorana. In Recami, E. et al. (Eds.). *Ettore Majorana —Nel centenario della sua nascita.* Catania: Edizioni Novecento.

Majorana, Q. (1938). Ulteriori ricerche sull'azione della luce su sottili lamine metalliche. *Il Nuovo Cimento, 5,* 573.

Mantegna, R. N. (2006). Majorana's article on 'The value of statistical laws in physics and social sciences'. *Proceedings of the Conference "Ettore Majorana's Legacy and the Physics of the XXI century",* PoS(EMC2006)011.

Mantegna, R. N., & Stanley, H. E. (1997). Physics investigation of financial markets. *Proceedings of the International School of Physics "Enrico Fermi",* Course CXXXIV, Società Italiana di Fisica, Bologna.

Mantegna, R. N., & Stanley, H. E. (1999). *Introduction to Econophysics: Correlations and complexity in finance.* Cambridge: Cambridge University Press.

Marinaro, M., & Scarpetta, G. (1996). *Eduardo R. Caianiello, Società Nazionale di Scienze.* Naples: Lettere e Arti.

Milton, K. A. (Ed.). (2000). *A Quantum legacy—seminal papers of Julian Schwinger.* Singapore: World Scientific.

Nastasi, P., & Tazzioli, R. (2005). Toward a scientific and personal biography of Tullio Levi-Civita (1873–1941). *Historia Mathematica, 32*(2), 203.

Orzalesi, C. A. (1968). Technical Paper No. 792. University of Maryland, Department of Physics and Astronomy.

Parravano, N. (1936). Il Fascismo e la Scienza. *Chimica e Industria, XVIII,* 222.

Pauli, W. (1941). Relativistic field theories of elementary particles. *Reviews of Modern Physics, 13,* 203.

Pauli, W. (1985). *Wissenschaften Briefwechsel mit Bohr, Einstein, Heisenberg, U.A. II: 1930– 1939.* Berlin: Springer.

Pauli, W. (1993). *Wissenschaften Briefwechsel mit Bohr, Einstein, Heisenberg, U.A. III: 1940– 1949.* Berlin: Springer.

Pauli, W., & Weisskopf, V. (1934). Über die Quantisierung der skalaren relativistischen Wellengleichung. *Helvetica Physica Acta, 7,* 709.

Pauling, L. (1931). The nature of the chemical bond. II. The one-electron bond and the three-electron bond. *Journal of the American Chemical Society, 53,* 3225.

Pauling, L. (1933). The normal state of the helium molecule-ions He_2^+ and He_2^{++}. *The Journal of Chemical Physics, 1,* 56.

Penrose, R. (2000). On Bell non-locality without probabilities: Some curious geometry. In Ellis, J. & Amati D. (Eds.). *Quantum Reflections.* Cambridge University Press, Cambridge.

Persico, E. (1936). *Fondamenti della meccanica atomica.* Bologna: Zanichelli.

Poggio, P. (1972). Il giovane fisico siciliano che morì per non vedere l'atomica. *Gente,* May 6.

Pontecorvo, B. (1958a). Mesonium and anti-mesonium. *Soviet Physics JETP, 6,* 429.

Pontecorvo, B. (1958b). Inverse β-processes and non-conservation of lepton charge. *Soviet Physics JETP, 7,* 172.

Preziosi, B. (Ed.). (1998). *Ettore Majorana: Lezioni all'Università di Napoli.* Naples: Bibliopolis.

Radicati di Brozolo, L. A. (1981). Antonio Carrelli. *Rendiconti dell'Accademia Nazionale dei Lincei, LXXI,* 243.

Randazzo, G. (1972). Ettore Majorana si uccise per una tragedia familiare. *Gente,* May 20.

Rasetti, F. (1936). *Il nucleo atomico.* Bologna: Zanichelli.

Recami, E. (1987). *Il caso Majorana—Epistolario, Documenti, Testimonianze.* Milan: Mondadori. Fourth revised and enlarged edition in Recami E. (2011). *Il caso Majorana—Epistolario, Documenti, Testimonianze.* Rome: Di Renzo.

Read, N., & Green, D. (2000). Paired states of fermions in two dimensions with breaking of parity and time-reversal symmetries and the fractional quantum Hall effect. *The Physical Review, B 61,* 10267.

Rivera, C., Infante, C., & Claro, F. (1967). Effective mass concept in solids: A classical analogy. *American Journal of Physics, 35,* 1143.

Roncoroni, S. (2011). Il promemoria 'Tunisi': un nuovo tassello del caso Majorana. *Il Nuovo Saggiatore, 27,* 58.

Roncoroni, S. (2012). Genesi dell'articolo postumo di Ettore Majorana. *Nuova Storia Contemporanea, 16,* 105.

Roth, L. (1963). Francesco Severi. *Journal of the London Mathematical Society, 38,* 282.

Russo, B. (1997). *Ettore Majorana—Un giorno di marzo.* Palermo: Flaccovio.

Sagnotti, A. (2013). Notes on strings and higher spins. *Journal of Physics A: Mathematical and Theoretical, 46,* 214006.

Schwinger, J. (1937). On Nonadiabatic processes in inhomogeneous fields. *The Physical Review, 51,* 648.

Schwinger, J. (1977). The Majorana formula. *Transactions of the New York Academy of Sciences, 38,* 170.

Segrè, E. (1960). Fermi, Enrico. In *Dizionario Biografico degli Italiani.* Rome: Treccani.

Segrè, E. (1970). *Enrico Fermi, physicist.* Chicago: The University of Chicago Press.

Segrè, E. (1993). *A mind always in motion.* Berkeley: University of California Press.

Severi, F. (1931). *Conferencias.* Buenos Aires: University of Buenos Aires.

Seyler, R. G., & Blanchard, C.H. (1961). Classical self-consistent nuclear model. *The Physical Review, 124,* 227.

Shenstone, A. G. (1931), Ultra-ionization potentials in mercury vapor. *The Physical Review, 38,* 873.

Speiser, A. (1927). *Theorie der Gruppen von Endlicher Ordnung.* Berlin: Springer.

Stückelberg, E. C. G. (1932). Theorie der unelastischen Stösse zwischen Atomen. *Helvetica Physica Acta, 5,* 369.

Toma, P. A. (2004). *Renato Caccioppoli—L'enigma.* Naples: Edizioni Scientifiche Italiane.

Tsvelik, A. M. (1990). Field-theory treatment of the Heisenberg spin-1 chain. *The Physical Review, 42,* 10499.

Ventimiglia, S. (2010). È così che si muore… in ricordo di Dorina Corso Majorana. *Sicilian Secrets.* http://blog.siciliansecrets.it/2014/08/29/e-cosi-che-si-muore-in-ricordo-di-dorina-corso-majorana/.

Vizioli, F. (1948). Michele Sciuti. *Rivista Sperimentale di Freniatria, 72*(3), 1.

Yukawa, H. (1935). On the interaction of elementary particles. I. *Nippon Sugaku-Buturigakkwai Kizi Dai 3 Ki, 17*, 48.

Weizel, W. (1930). Molekülzustände des Wasserstoffs mit zwei angeregten Elektronen. *Zeitschrift für Physik, 65*, 456.

Wentzel, G. (1927). Über strahlungslose Quantensprünge. *Zeitschrift für Physik, 43*, 524.

Wentzel, G. (1943). *Einführung in die Quantentheorie der Wellenfelder.* Vienna: Deuticke.

Weisskopf, V. (1973). My life as a physicist. In Zichichi A. (Ed.), *Properties of the fundamental interactions.* Bologna: Editrice Compositori.

Weyl, H. (1928). *Gruppentheorie und Quantenmechanik.* Leipzig: Hirzel.

Wick, G. C. (1981). Fisica a Roma negli anni '30 e '40. *Atti dell'Accademia delle Scienze di Torino, Supplement 2, 115*, 13. See also Wick, G. C. (1988). Physics and Physicists in the thirties. In Zichichi A. (Ed.). *Proceedings of the 22nd Course of the International School of Subnuclear Physics.* New York: Plenum Press.

Wigner, E. (1939). On unitary representations of the inhomogeneous Lorentz group. *Annals of Mathematics, 40*, 149.

Wigner, E. (1948). Relativistische Wellengleichungen. *Zeitschrift für Physik, 124*, 665.

Wilczek, F. (2009). Majorana returns. *Nature Physics, 5*, 614.

Wu, T.-Y. (1934). Energy states of doubly excited helium. *The Physical Review, 46*, 239.

Wu, T.-Y. (1944). Auto-ionization in doubly excited helium and the λ320.4 and λ357.5 Lines. *The Physical Review, 66*, 291.

Zener, C. (1932). Non-Adiabatic crossing of energy levels. *Proceedings of the Royal Society of London, A 137*, 696.

Zullino, P. (1964). Chi ha visto il genio senza volto? *Epoca,* June 14.

Bibliography

Articles published by Majorana

P1. Gentile, G., & Majorana, E. (1928). Sullo sdoppiamento dei termini Roentgen ottici a causa dell'elettrone rotante e sulla intensità delle righe del Cesio ("On the splitting of the Roentgen and optical terms caused by the spinning electron and on the intensity of the caesium lines"). *Rendiconti dell'Accademia dei Lincei, 8*, 229–233.

P2. Majorana, E. (1931). Sulla formazione dello ione molecolare di He ("On the formation of the helium molecular ion"). *Il Nuovo Cimento, 8*, 22–28.

P3. Majorana, E. (1931). I presunti termini anomali dell'Elio ("On the presumed anomalous terms of helium"). *Il Nuovo Cimento, 8*, 78–83.

P4. Majorana, E. (1931). Reazione pseudopolare fra atomi di Idrogeno ("Pseudopolar reaction of hydrogen atoms"). *Rendiconti dell'Accademia dei Lincei, 13*, 58–62.

P5. Majorana, E. (1931). Teoria dei tripletti P' incompleti ("Theory of the incomplete P' triplets"). *Il Nuovo Cimento, 8*, 107–113.

P6. Majorana, E. (1932). Atomi orientati in campo magnetico variabile ("Oriented atoms in a variable magnetic field"). *Il Nuovo Cimento, 9*, 43–50.

P7. Majorana, E. (1932). Teoria relativistica di particelle con momento intrinseco arbitrario ("Relativistic theory of particles with arbitrary intrinsic angular momentum"). *Il Nuovo Cimento, 9*, 335–344.

P8a. Majorana, E. (1933). Uber die Kerntheorie ("On nuclear theory"). *Zeitschrift für Physik, 82*, 137–145.

P8b. Majorana, E. (1933). Sulla teoria dei nuclei ("On nuclear theory"). *La Ricerca Scientifica, 4*, 559–565.

P9. Majorana, E. (1937). Teoria simmetrica dell'elettrone e del positrone ("Symmetric theory of electrons and positrons"). *Il Nuovo Cimento, 14*, 171–184.

P10. Majorana, E. (1942). Il valore delle leggi statistiche nella fisica e nelle scienze sociali ("The value of statistical laws in physics and social sciences"). *Scientia, 36*, 58–66.

Majorana's Volumetti

Esposito, S., Majorana, E. Jr., van der Merwe, A., & Recami, E. (2003). *Ettore Majorana—Notes on theoretical physics*. Dordrecht: Kluwer Academic Publishers.

Esposito, S., & Recami, E. (2006). *Ettore Majorana—Appunti inediti di Fisica teorica*. Bologna: Zanichelli.

Majorana's Quaderni

Esposito, S., Recami, E., van der Merwe, A., & Battiston, R. (2008). *Ettore Majorana—Unpublished research notes on theoretical physics*. Heidelberg: Springer.

Majorana's lectures at the University of Naples

© Springer International Publishing AG 2017
S. Esposito, *Ettore Majorana*, Springer Biographies,
DOI 10.1007/978-3-319-54319-2

Esposito, S. (Ed.) (2006). *Ettore Majorana—Lezioni di Fisica Teorica*. Naples: Bibliopolis.
Preziosi, B. (Ed.) (1987). *Ettore Majorana—Lezioni all'Università di Napoli*. Naples: Bibliopolis.

Studies on (unpublished) manuscripts by Majorana

Di Grezia, E. (2006). Majorana and the investigation of infrared spectra of ammonia. *Electronic Journal of Theoretical Physics, 3*, 225–238.
Di Grezia, E., & Esposito, S. (2008). Majorana and the quasi-stationary states in nuclear physics. *Foundations of Physics, 38*, 228–240.
Di Mauro, M., Esposito, S., & Naddeo, A. (2016). Majorana and the theoretical problem of photon-electron scattering. *Advances in Historical Studies, 5*, 113–125.
Esposito, S. (2002). Majorana solution of the Thomas-Fermi equation. *American Journal of Physics, 70*, 852–856.
Esposito, S. (2006). Majorana and the path-integral approach to Quantum Mechanics. *Annales de la Fondation Louis de Broglie, 31*, 207–225.
Esposito, S. (2006). A peculiar lecture by Ettore Majorana. *European Journal of Physics, 27*, 1147–1156.
Esposito, S. (2007). Hole theory and Quantum Electrodynamics in an unknown manuscript in French by Ettore Majorana. *Foundations of Physics, 37*, 956–976 (also at 1049–1068).
Esposito, S. (2007). An unknown story: Majorana and the Pauli-Weisskopf scalar electrodynamics. *Annalen der Physik* (Leipzig), *16*, 824–841.
Esposito, S. (2009). A theory of ferromagnetism by Ettore Majorana. *Annals of Physics* (New York), *324*, 16–29.
Esposito, S. (2012). Searching for an equation: Dirac, Majorana and the others. *Annals of Physics* (New York), *327*, 1617–1644.
Esposito, S. (2014). *The physics of Ettore Majorana: phenomenological, theoretical and mathematical*. Cambridge: Cambridge University Press.
Esposito, S., & Naddeo, A. (2012). Majorana solutions to the two-electron problem. *Foundations of Physics, 42*, 1586–1608.
Esposito, S., & Naddeo, A. (2015). Homopolar bond and ionic structures: Two contributions by Majorana. *Annales de la Fondation Louis de Broglie, 40*, 157–179.
Esposito, S., & Naddeo, A. (2015). Majorana, Pauling and the quantum theory of the chemical bond. *Annalen der Physik* (Berlin), *527*, A29–A33.
Esposito, S., & Salesi, G. (2010). Fundamental times, lengths and physical constants: Some unknown contributions by Ettore Majorana. *Annalen der Physik* (Berlin), *522*, 456–466.

Studies related to Majorana's scientific work

Drago, A., & Esposito, S. (2006). A logical analysis of Majorana's papers on theoretical physics. *Electronic Journal of Theoretical Physics, 3*, 249–263.
Esposito, S. (1998). Covariant Majorana formulation of Electrodynamics. *Foundations of Physics, 28*, 231–244.
Esposito, S. (2002). Majorana transformation for differential equations. *International Journal of Theoretical Physics, 41*, 2417–2426.
Esposito, S. (2005). Again on Majorana and the Thomas-Fermi model: A comment to physics/0511222, e-print arXiv:physics/0512259.
Esposito, S. (2006). Four variations on theoretical physics by Ettore Majorana. *Electronic Journal of Theoretical Physics, 3*, 265–283.
Fradkin, D. (1966). Comments on a paper by Majorana concerning elementary particles. *American Journal of Physics, 34*, 314–318.
Giannetto, E. (1985). A Majorana-Oppenheimer formulation of Quantum Electrodynamics. *Lettere al Nuovo Cimento, 44* (1985), 140–144.
Giannetto, E. (1985). A Majorana-Dirac-like equation for a non-abelian gauge field. *Lettere al Nuovo Cimento, 44* (1985), 145–148.

Leonardi, C., Lillo, F., Vaglica, A., & Vetri, G. (1999). Majorana and Fano alternatives to the Hilbert space. In Bonifacio R. (Ed.). *Mysteries, puzzles, and paradoxes in quantum mechanics*. Woodbury, N.Y.: AIP.

Mignani, R., Baldo, M., & Recami, E. (1974). About a Dirac-like equation for the photon, according to Ettore Majorana. *Lettere al Nuovo Cimento, 11*, 568–575.

Plyushchay, M. S. (2006). Majorana equation and exotics: Higher derivative models, anyons and noncommutative geometry. *Electronic Journal of Theoretical Physics, 3*, 17–31.

Studies on the historical and scientific context of Majorana's work

Arimondo, E., Clark, C., & Martin, W. (2010). Ettore Majorana and the birth of autoionization. *Reviews of Modern Physics, 82*, 1947–1958.

De Gregorio, A. (2007). Il 'protone neutro', ovvero della laboriosa esclusione degli elettroni dal nucleo. *Physis, 44*, 153–184.

De Gregorio, A., & Esposito, S. (2006). A lezione dal Professor Majorana. *Sapere, 72*(3), 56–60.

De Gregorio, A., & Esposito, S. (2007). Teaching theoretical physics: The cases of Enrico Fermi and Ettore Majorana. *American Journal of Physics, 75*, 781–790.

Di Grezia, E., & Esposito, S. (2004). Fermi, Majorana and the statistical model of atoms. *Foundations of Physics, 34*, 1431–1450.

Drago, A., & Esposito, S. (2004). Following Weyl on Quantum Mechanics: The contribution of Ettore Majorana. *Foundations of Physics, 34*, 871–887.

Drago, A., & Esposito, S. (2007). Ettore Majorana's course on Theoretical Physics: a recent discovery. *Physics in Perspective, 9*, 329–345.

Esposito, S. (2005). Il corso di Fisica teorica di Ettore Majorana: il ritrovamento del Documento Moreno. *Il Nuovo Saggiatore, 21*, 21–33.

Esposito, S. (2008). Ettore Majorana and his heritage seventy years later. *Annalen der Physik* (Leipzig), *17*, 302–318.

Esposito, S., & Recami, E. (2003). The Volumetti by Ettore Majorana. In Gariboldi L., & Tucci P. (Eds.). *History of Physics and Astronomy in Italy in the 19th and 20th centuries: Sources, themes and international context*. Milan: s.n.t.

Giannetto, E. (1988). Su alcuni manoscritti inediti di E. Majorana. In F. Bevilacqua (Ed.). *Atti del IX Congresso Nazionale di Storia della Fisica*. Milan: s.n.t.

Giannetto, E. (1993). Ettore Majorana e gli stati ad energia negativa. *Giornale di Fisica, 34*, 151–158.

Recami, E. (1999). Ricordo di Ettore Majorana a sessant'anni dalla sua scomparsa: l'opera scientifica edita e inedita. *Quaderni di Storia della Fisica, 5*, 19–68.

Roncoroni, S. (2012). Genesi dell'articolo postumo di Ettore Majorana. *Nuova Storia Contemporanea, 16*, 105.

The following bibliographic references are not meant to complete the many works about Ettore Majorana, but merely to give a better overall picture of the Sicilian physicist's life and scientific work.

Biographical works

Amaldi, E. (1966). *La vita e l'opera di E. Majorana*. Accademia Nazionale dei Lincei, Rome. English translation: Amaldi, E. (1966). *Ettore Majorana: Man and scientist*. In Zichichi A. (Ed.). *Strong and weak interactions*. New York: Academic Press.

Amaldi, E. (1968). Ricordo di Ettore Majorana. *Giornale di fisica, 9*, 300–309.

Amaldi, E. (1988). Ettore Majorana, a cinquant'anni dalla sua scomparsa. *Il Nuovo Saggiatore, 4*, 13–26.

Esposito, S. (2006). Fleeting genius. *Physics World, 19*, 34–36.

Esposito, S. (2009). *La cattedra vacante*. Naples: Liguori.

Esposito, S. (2010). The disappearance of Ettore Majorana. *Contemporary Physics, 51*, 193–209.

Recami, E. (1987). *Il caso Majorana—Epistolario, Documenti, Testimonianze*. Milan: Mondadori.
Fourth revised and enlarged edition in Recami, E. (2011). *Il caso Majorana—Epistolario,*
Documenti, Testimonianze. Rome: Di Renzo.
Russo, B. (1997). *Ettore Majorana—Un giorno di marzo*. Palermo: Flaccovio.

Other historical and biographical works

Amaldi, E. (1984). From the discovery of the neutron to the discovery of nuclear fission. *Physics*
Reports, 111, 1–331.
Cordella, F., De Gregorio, A., & Sebastiani, F. (2001). *Enrico Fermi. Gli anni italiani*. Rome:
Editori Riuniti.
Esposito, S. (2010). The Majorana mystery. *Physics World, 23*, 44–45.
Fermi, E. (1937). Un maestro: O. M. Corbino. *Nuova Antologia, 72*, 313–316.
Fermi, L. (1954). *Atoms in the family*. Chicago: The University of Chicago Press.
Gentile, B. (1988). Lettere inedite di E. Majorana a G. Gentile Jr. *Giornale critico della filosofia*
italiana, 8, 145–153.
Macorini, E. (Ed.) (1974). *Scienziati e tecnologi contemporanei: Enciclopedia Biografica. Volume*
3. Milan: Mondadori.
Pontecorvo, B. (1972). *Fermi e la fisica moderna*. Rome: Editori Riuniti.
Roncoroni, S. (2013). *Ettore Majorana, lo scomparso e la decisione irrevocabile*. Rome: Editori
Internazionali Riuniti.
Segrè, E. (1970). *Enrico Fermi, physicist*. Chicago: The University of Chicago Press.
Segrè, E. (1993). *A mind always in motion*. Berkeley: University of California Press.
Toma, P. A. (2004). *Renato Caccioppoli—L'enigma*. Naples: Edizioni Scientifiche Italiane.

Other useful reference works

Dizionario Biografico degli Italiani. Rome: Treccani. 1960ff.
Bethe, H. A. (1955). Memorial symposium held in honor of Enrico Fermi at the Washington
meeting of the American Physical Society, April 29. *Reviews of Modern Physics, 27*, 249ff.
Castellani, L. (1974). *Dossier Majorana*. Milan: Fabbri.
Castelnuovo, G. (1956). *Lezioni di geometria analitica*. Rome: Società Editrice Dante Alighieri.
Fermi, E. (1960). *Collected Papers*. Rome-Chicago: Accademia Nazionale dei Lincei—The
University of Chicago Press.
Goodstein, J. R. (2007). *The Volterra Chronicle: The life and times of an extraordinary*
mathematician 1860–1940. Providence-London: American Mathematical Society—London
Mathematical Society.
Levi-Civita, T., & Amaldi, U. (1923). *Lezioni di Meccanica Razionale*. Bologna: Zanichelli.
Nastasi, P., & Tazzioli, R. (2005). Toward a scientific and personal biography of Tullio
Levi-Civita (1873–1941). *Historia Mathematica, 32*(2), 203–236.
Radicati di Brozolo, L. A. (1981). Antonio Carrelli. *Rendiconti dell'Accademia del Lincei, LXXI*,
243.
Rasetti, F. Unpublished autobiographical note, kept at the Amaldi Archive of the Department of
Physics of Sapienza University of Rome.
Roth, L. (1963). Francesco Severi. *Journal of the London Mathematical Society, 38*, 282–307.

Timeline

1906

August 5

- Ettore Majorana is born in Catania, fourth of the five sons of Fabio Massimo (1875–1934) and Salvatrice (Dorina) Corso (1876–1965).

1921

- The Majoranas move to Rome.

1923

July

- He gets his standard high school certificate at the State *Liceo-Ginnasio "Torquato Tasso"* in Rome.

November 3

- He enrols at the *Biennio di Studi di Ingegneria* (initial two-year engineering course) of the University of Rome.

1925

December 3

- He enrols at the *Scuola di Applicazione degli Ingegneri* (final three-year course of the School of Engineering).

1927

- He starts writing the *Volumetti*, copybooks of personal notes where he records some of his studies and/or research (the date on the first of the five copybooks is March 8, 1927).

Summer

- E. Amaldi and E. Segrè accept O.M. Corbino's plea and accept to change from engineering to physics (this move will be officialised in November 1927 for Amaldi and February 8, 1928 for Segrè).

© Springer International Publishing AG 2017
S. Esposito, *Ettore Majorana*, Springer Biographies,
DOI 10.1007/978-3-319-54319-2

Autumn

- Segrè convinces Ettore to meet E. Fermi; after this meeting he will change to physics, too (the official move is dated November 9, 1928).

1928

- Together with G. Gentile, he publishes his first paper on the area of spectroscopic research studied by the Fermi group: *On the splitting of the Roentgen and optical terms caused by the spinning electron and on the intensity of the caesium lines* (the paper is presented at the *Accademia dei Lincei* on July 24, and published in its *Rendiconti*).

December 29

- Still a university student, he attends the XXII General Meeting of the Italian Physical Society (chaired by his uncle Quirino) and presents a scientific note, *Search for a general expression of Rydberg corrections, valid for neutral atoms or positive ions* (the report can be found in the journal *Il Nuovo Cimento*, 6, XIV–XVI, while the original study is reported in *Volumetto* no. II).

1929

July 6

- He graduates in physics with honours, presenting the thesis *Quantum theory of radioactive nuclei*. It is the first time a topic relating to nuclear physics is studied in the *Via Panisperna* group in Rome.

1931

- He publishes two papers on the chemical bond in molecules, *On the formation of the helium molecular ion* (in *Il Nuovo Cimento*) and *Pseudopolar reaction of hydrogen atoms* (presented at the *Accademia dei Lincei* on January 4 and published in its *Rendiconti*), and another two papers on spectroscopic research, *On the presumed anomalous terms of helium* and *Theory of the incomplete P' triplets* (both in *Il Nuovo Cimento*).

1932

- He publishes two more papers in *Il Nuovo Cimento*: *Oriented atoms in a variable magnetic field* (solicited by Segrè) and *Relativistic theory of particles with arbitrary intrinsic angular momentum*.

March

- After Chadwick's announcement of the discovery of the neutron, Majorana tells the other members of the *Via Panisperna* institute that he is working on a theory of light nuclei comprising only protons and neutrons; though encouraged by Fermi and his group, Ettore decides he will not publish his work. A similar theory, with some imperfections, will be published independently by W. Heisenberg in the following July.

November 12

- He obtains a lecturing post (*libera docenza*) in theoretical physics at the University of Rome (the administrative order is dated January 12, 1933).

1933

January

- On Fermi's proposal to the C.N.R., Ettore is granted a scholarship to study abroad.

January 19

- Ettore arrives in Leipzig in the evening to work at the Institute of Theoretical Physics directed by Heisenberg.

March 3

- Encouraged by Heisenberg, he sends the German journal *Zeitschrift für Physik* the paper *Über die Kerntheorie* (On nuclear theory), whose main content was already ready a year earlier, and where Heisenberg's theory of nuclear forces is corrected.

March 4

- From Leipzig he goes to Copenhagen and stays there for about a month at the Institute of Theoretical Physics directed by Niels Bohr.

April 15

- Ettore leaves Copenhagen to spend the Easter holidays in Rome.

May

- At the University of Rome, he presents the programme of the open course on *Mathematical methods of quantum mechanics*, never actually given.

May 11

- On C.N.R.'s request, he sends the Italian version of the German paper *Über die Kerntheorie*, to the periodical *La Ricerca Scientifica* published by this institution.

July

- The Majoranas (his mother, his two sisters Rosina and Maria, and his brother Salvatore) pay a visit to Ettore in Leipzig.

July 31

- He is enrolled in the Fascist Party (we do not know who filed the request).

August 5

- Ettore goes back to Rome, after his stay in Germany.

1934

July 11

- Ettore's father dies in Rome; this event will have a very negative influence on the Sicilian physicist's life.

1935

April 30

- At the University of Rome, he presents the programme for the open course *Mathematical methods of atomic physics*, never actually given.

1936

March

- Invited by his uncle Giuseppe, he sends him the final draft of the manuscript *The value of statistical laws in physics and social sciences*. It will be published posthumously by G. Gentile in 1942 (in the journal *Scientia*).

April 28

- At the University of Rome, he presents the programme of the open course on *Quantum electrodynamics*, never actually given.

1937

- The article *Symmetric theory of electrons and positrons* is published in *Il Nuovo Cimento*, where Ettore presents a study carried out some time earlier. It contains the fundamental theory of *Majorana's neutrino*.

June 15

- Deadline to enter the general selection for the tenure of theoretical physics at the University of Palermo (requested by Segrè); Ettore is among the candidates (besides G. Gentile, L. Pincherle, G. Racah, G. Wataghin, and G.C. Wick).

October 25

- The examining board for the tenure of theoretical physics (chaired by Fermi) meet for the first time, but they soon interrupt their deliberations to send a letter to the Minister of Education, G. Bottai, with a request to give a tenure to Majorana outside the rules of the selection, "for his high and well deserved repute".

November 2

- The Minister Bottai accepts the board's request and appoints Majorana, to start from the following November 16.
- Ettore begins an epistolary exchange with A. Carrelli, director of the Institute of Physics at the University of Naples, where he will hold his tenure of theoretical physics.

December 4

- The administrative order is registered at the Court of Auditors and becomes effective; in the following days, the appointment is communicated to Ettore at his residence in Rome.

1938

January 10 or 11

- Ettore arrives in Naples to take up his tenure of theoretical physics.

January 13

- At 9 o'clock he gives the opening lecture of the theoretical physics course. His family arrives from Rome to attend the event.

January 15

- He actually begins the course with his students, to be given on even week days (except holidays).

January 17

- Ettore make his oath "to the King and the Fascist Regime" before the chancellor G. Salvi, as required of all university professors.

January 18

- He meets the experimental physicist G. Occhialini, back from South America, who is paying a visit to the director Carrelli.

March 12

- After giving his lecture, Ettore goes to Rome to visit his family for the last time.

March 24

- He gives his last lecture (no. 21) to the students.

March 25

- In the morning he goes to the Institute of Physics to give one of his students the folder with his lecture notes. After going back to the hotel, at about 5 o'clock in the afternoon, he leaves it for an unknown destination. Here end the (confirmed) traces of Majorana's ensuing fate.

Printed in the United States
By Bookmasters